The Structure of Matter

The Growth of Man's Ideas on the Nature of Matter

Physics and Humanities Series

The Structure of Space: *The Growth of Man's Ideas on the Nature of Forces, Fields and Waves*
by Joan Solomon

A HALSTED PRESS BOOK

The Structure
of
Matter

*The Growth of Man's Ideas on the
Nature of Matter*

Joan Solomon

JOHN WILEY & SONS NEW YORK

ISBN 0470–81222–2

Library of Congress Catalog Card Number 73–8541

Published in the U.S.A.
by Halsted Press, a Division
of John Wiley & Sons, Inc.
New York

Printed in Great Britain

Contents

5

CONTENTS

Introduction

Those who feel a diffident wish to learn a little more about Science may well be discouraged by a host of difficulties. They may fear that the field is too vast, that the language is incomprehensible, and that the whole approach is alien. I believe that none of these obstacles is real. In the first instance it is valuable to confine one's reading to the growth of ideas on a single topic, like the nature of substances; secondly the technical language of mathematics is not necessary: it has always been rendered into clear, even beautiful words by the great among scientists; and thirdly the 'culture' of science, if such exists, has always been reassuringly familiar.

Reading the original words of scientists past and present introduces us unmistakably to men who shared the passions, fears, and beliefs peculiar to their times. Each period in history tends to think and deliberate in a characteristic fashion which links together its Science, Literature, Art, and Philosophy, and this survey, however scanty, should give some inkling of these trends of thought. It is quite possible to pick and choose among the sections in this book those times or ways of thinking that appeal most to the reader or it may be read as a continuous story in which each age added its own characteristic contribution to the elucidation of matter.

History apart, this approach may have other benefits to offer. The student of science and the student of humanities should erect no barriers and level no taunts of 'illiterate' or 'uncultured'. The heroes of science were always a part of the mainstream of contemporary endeavour, and the valued friends of those poets and philosophers that the 'culture-fiends' still revere so much! But then there never have been two cultures; we are all creatures of our total thinking environment and by observing how our generation of scientists struggles to answer the perennial queries about the stuff of substance we may even learn to recognise a little more about the characteristics of our own times.

I
Myth and Speculation

This is a series of glimpses at some of the scientific theories that men have held about the *matter* of which this world is made. Nature is very rich in different materials from fluids and crystals to green shoots and living tissue yet, for all their apparent differences, these comprise the whole body of tangible substances of our world. More 'real' than the unseen forces behind thunderstorms, gravity, or earthquakes, the natural materials seem to be the most obvious starting point for science—for a kind of primitive chemistry—when men should first ask 'What are these things made from?' But the very first questions were not asked in quite that way and the first answers were simple stories and myths about the creation of the world and the life it holds. It was out of this treasury of legends that the first recognisable scientific theories arose.

The primitive Aborigines of Australia, who live now as our own ancestors must have lived some 20,000 years ago, have their myths about the creation of the world in the beginning of time which they call 'the dreaming'. It was then, they say, that Eingama the Snake, whom they call Mother, rose from the water with all life in her belly and spewed out men, birds, and animals across the empty land. These original Australians, who still live in an Old Stone Age culture with only wooden spears and boomerangs, know so little about the substances that can be coaxed out of Nature that the simplicity of their creation story must seem a poor starting point for science. More sophisticated peoples learnt to melt metals from their ores, bake hard their clay pots, and mix subtle glazes from special powdered earths—yet their legends seem just as simple and naive.

There was a deep need for these myths among primitive peoples

Zeus and Athene fighting two of the Titans. In later legends the Titans were remembered only as bad giants whom the gods had defeated

living precariously, as they did, in a largely hostile world. This was the only way that they could reconcile themselves to their fate. The cause of disaster and death in the world of the Aborigines was man's first disobedience just as it was for Adam and Eve in our own bible when they were cast out of Eden into this hard world. If man believed that the bitterness of life was his own fault he could not, like Job, turn against God. The telling and retelling of these stories was also a religious act by which they hoped to propitiate a powerful spirit who might otherwise wreak yet more vengeance on hard-pressed man. So these myths were an active link between gods and men: conciliating, reconciling, and communing. When scientific theories emerged they were clearly different because their exponents did not expect the forces of Nature to react vengefully—nor indeed in any other way!—to what they chose to assert. A theory was *wrong* only because it 'did not work', not because it was displeasing to the gods. The earliest scientists may well have believed in a Creator but they did not look back to the moment of creation but forward through the power of their speculation. They did not seek to bring comfort and illusion to the soul of man as the life-giving myth had done. Science was the child of unrestrained imagination and intellectual wonder and, though its earliest endeavours were little different in content from the myths that had gone before, their purpose was very different.

The Greek Primeval Gods

One of the earliest fragments of Greek papyrus is an account of the origin of the universe attributed to Pherecydes of Scyros who lived about the seventh century BC. Legend connects him with both Pythagoras and Thales and this work gives a good idea of the current background of mythology, out of which these two scientific 'giants' of the ancient world framed their theories. In his book Zeus (Sun) and Chthonie (Earth) marry and they beget Eros who brings harmony into the world.

Their next children are the Titans: Fire, Breath, and Water. If this represents a typical legend of the time, it is easy to see where the great Ionic philosophers of the following century got the germs of their ideas.

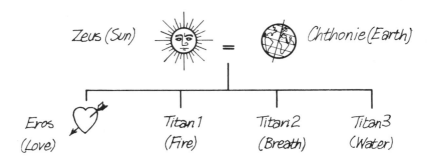

Zeus (Sun) = Chthonie (Earth)

| | | | |
| Eros (Love) | Titan 1 (Fire) | Titan 2 (Breath) | Titan 3 (Water) |

The Ionian Philosophers speculate

Thales was acknowledged as the first of the Greek philosophers. He lived in Miletus, a Greek town in Asia Minor, and although he left no written works, he can be dated by the fact that he predicted the year of a solar eclipse. This was recorded by the historian Herodotus as it put a dramatic end to the war between Lydia and Media, which had been dragging on for six years, by terrifying the combatants in mid-battle. The year was 585 BC.

Thales was said to have been well travelled, he had visited both Egypt, where he observed the vital yearly flooding of the Nile, and had lived in Mesopotamia where, no doubt, he had heard the legend of the great flood. Later Greek writers maintain that he was the first to propose a theory about the nature of the matter or substances of which the world is made. This is what Aristotle wrote about his theory some 200 years later in the *Metaphysics*:

> Most of the earliest philosophers thought that the principles which were in the nature of matter were the only principles of all things: that of which all things that are consist and from which they first came to be and into which they are resolved as a final state (the substance remaining but changing its modifications) this,

they said, is the element and principle of all things, and therefore they think that nothing is either generated or destroyed. . . . On the number and nature of such principles they do not all agree. Thales, who led the way in this kind of philosophy, says the principle is water, and for this reason declared that the earth rests on water.

The 'principle' to which Aristotle referred was called 'arché' by the early Greeks which suggests its primaeval character. Not only was it indestructible and eternal but it was also the unifying element which lay at the root of the myriads of different substances of which

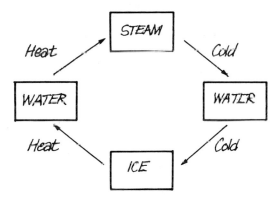

the world is composed. The attempt to simplify the complexities of all the materials around us had begun and was to continue, with some set-backs, until the present day. The idea of a cycle is also clear in what Aristotle wrote. What kind of cycle Thales was considering is a matter for conjecture but it does not seem to have been brought about by either magic or the hand of the gods. This is the reason why Thales can be taken as the starting point in the study of European science. The cycles, the changes, are natural. Through this way of looking at substances the early world came gradually to accept the unstated principle upon which chemistry stands or falls: MATTER CANNOT BE CREATED OR DESTROYED. This is known today as the Law of Conservation of Matter, and when it is clearly understood, when magical appearances and disappearances cease, then the study of substances can begin.

Anaximander was a fellow citizen of Thales and probably his pupil. He also believed in an 'arché', but for him this elementary material was boundless—so deeply buried beneath appearances as to be almost unknowable. Hardly a useful concept for the basis of a practical science! This was for him the undestroyable element from which the 'warring' substances arise and to which they return.

Anaximenes was the third and last of the great Milesian philosophers and a younger contemporary of Anaximander. He seems to have been a more down-to-earth thinker; not only did he postulate a real substance, air, as the primordial element or 'arché' but gave thought to how it could be changed by pressure, movement, or heat to give other substances. Aristotle quotes his theory in *Physics* (24.26.A5):

> Anaximenes . . . also posits a single, infinite, underlying substance of things, not, however, indefinite in character like Anaximander's but determinate, for he calls it air and says that it differs in rarity and density according to the different substances: Rarefied it becomes fire; condensed, it becomes first wind, then cloud, and when condensed still further water, then earth and stones. Everything else is made of these. He too postulated eternal motion, which is indeed the cause of the change.

Anaximenes was also prepared to back his theory by reference to actual experience. He is said to have given the example of the cold air that is blown out when you compress your lips in contrast to the warm air which is exhaled when your mouth is relaxed and open. This is to illustrate his theory of compression and rarefaction.

The ideas of the Milesian school spread throughout the Greek-speaking world. Many more speculations were made; some of these ancient philosophers favoured water as the primary element, some favoured fire. Few of them related their theories to practical examples and some even taught that observation was not to be trusted and that reality could only be apprehended in the mind. Others concentrated on the idea of movement and conceived of the world in terms of ceaseless whirls of tiny, indivisible, atoms spinning and colliding in a void. To Leucippus and Democritus belongs the credit for inventing this original atomic theory. Unfortunately it had few followers in the ancient world, partly because the founders were considered atheists, and partly because the great Aristotle and others poured scorn on it. Their criticism was that the sentence 'a void exists' is logically impossible since 'existence' presupposes *something* and a 'void' is *nothing*. This language difficulty was enough to damn the theory in the most influential circles of the Ancient World!

The Four Element Theory

The Greek theory of the elements was finally completed by Empedocles in about 450 BC. He taught his followers explicitly to use their senses as well as their minds to understand the Universe. One of the remaining fragments of his verse describes very charmingly his proof of the reality of air by observation and even measurement.

> As when a girl playing with a (funnel) of shining brass—when, having placed the mouth of the pipe on her well-shaped hand she dips the vessel into the yielding substance of water, still the volume of air pressing from the inside on the many holes keeps out the water, until she uncovers the condensed stream (of air). Then at once when the air flows out, the water flows in in an equal quantity.

It was this philosopher, Empedocles, who finally combined the four 'ache' or elements: FIRE, AIR, WATER, AND EARTH into one theory of matter. He imagined them ceaselessly uniting and then separating out again as they formed more complex substances which, in time, decomposed into the simple elements from which they had

15

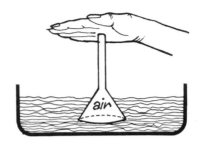

been made. Even animals and men, he taught, were no more than combinations of these elements.

> But men, when these [Elements] have been mixed in the form of a man and come into the light, or in the form of a species of wild animals, or plants, or birds, then say that this has 'come into being'; and when they separate, this men call sad fate (death).

This theory, with a few minor additions, was to stand unchallenged for two thousand years—an astonishing span of time—and was known all over Europe from scholars to the man-in-the-street. Even in Elizabethan times Shakespeare, echoing Empedocles himself, could make Antony say of Brutus [*Julius Caesar*]

> *His life was gentle, and the elements*
> *So mix'd in him that Nature might stand up*
> *And say to all the world, 'This was a man!'*

Empedocles represents, perhaps, the high-spot in Greek science. Not only did he assemble the four-element theory but he is credited with several more experiments with inflated wine-skins. Men said that he was trying to control the winds and that he thought himself a god. There is even a legend that he jumped into the active crater of Mount Etna to prove himself immortal—definitely a colourful character!

2
Philosophical Disaster

From 428 to 348 BC there lived in Athens the philosopher whose whole way of thought was to light the civilised world for two thousand years. This was Plato, friend and pupil of the ill-fated Socrates, and founder of the great Academy—the first university of the west. Plato sought in his philosophy to make clear the eternal pattern that lay behind all the changing appearances of this world. Such a scheme, he believed, should be as elegant and timeless as the propositions of mathematics to be worthy of the divine mind of a creator. Fashioned by pure intelligence, the universe was to be understood only by intelligence, and since all nature had been endowed with 'invisible soul' (the power of change and movement) it should be observed only by the soul of man. In his dialogue *Phaedo* he puts these words into the mouth of Socrates: 'If we are ever to know anything absolutely, we must be free of the body and behold the actual realities with the eye of the soul alone.' Upon this principle Plato built his philosophy—speculative, inward-looking, and remote from the world of touch and sight.

What is real?
Plato used the scientific theories of his time, but not as a scientist. He liked Anaximander's picture of a single, eternal and invisible 'arché', the underlying stuff of matter, and held that it was always unchanged although it took on the qualities of Empedocles' four elements. It could *seem* to resemble fire, air, water, or earth but its essential nature never varied. This intangible ghost-fabric alone was permanent though it 'appears to possess different qualities at different times; while things that pass in and out of it are copies of eternal

things' [*Timaeus*]. These 'eternal things' were the models or true forms of the four elements which existed before the heavens or the earth were made and which could be perceived beneath the trivial presence of actual fire, air, water, or earth, by the use of reason alone.

Real science is concerned with perceptible objects and their per-

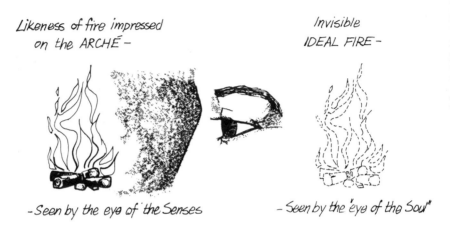

Likeness of fire impressed
on the ARCHÉ –

Invisible
IDEAL FIRE –

– Seen by the eye of the Senses

– Seen by the 'eye of the Soul"

ceptible movement—but Plato taught that it was a nobler task to contemplate the eternal truths that lie behind the visible façade of nature. In his book *Timaeus* he does go so far as to discuss the actual appearance of some 'watery' metals and to expound, by way of light relief, a little theoretical chemistry :

> Of all these fusible varieties of water, as we have called them, one is gold. It is very dense, consisting of extremely fine and uniform particles. . . .
>
> Another variety has particles closely resembling those of gold, but of more than one grade. . . . This formation is copper, one of the bright and solid kinds of water. The portion of earth mixed with it appears by itself on the surface when the two substances begin to be separated again by the action of time; it is called 'verdigris'.
>
> To enumerate the other substances of this kind . . . would be no complicated task. When a man, by way of recreation, lays aside

discourse about eternal realities and derives harmless amusement from such plausible accounts of becoming, he will make for his life a sober and sensible pastime.

This condescending attitude towards science did little to recommend such 'harmless amusement' to his pupils! Time was only to increase Plato's reputation and influence. Later, the early Christian church invested his work with almost divine significance because it could so easily accommodate an eternal, perfect, and all-knowing God. Similarly the imperfections and injustices of this world could be safely ignored as poor copies while man, through prayer, meditated on the God-made ideals. Neo-Platonic Christian sects abounded in the early centuries of the new era but such devout acceptance of Plato's teaching was to prove almost disastrous to the growth of science.

Qualities not elements

Plato's most famous pupil was Aristotle. He was the son of a doctor and in some ways was a more practical man than Plato—he certainly had a deep and life-long interest in biology. However, when it came to the basic nature of matter Aristotle generally followed Plato's lead. He named four basic qualities: 'hot', 'cold', 'moist', and 'dry'; and combined them in pairs to form the same elements that Empedocles had postulated a century earlier.

> The elementary qualities are four . . . it is evident that the 'couplings' of the elementary qualities will be four: hot with dry and moist with hot, and again cold with dry and cold with moist. And these four couples have attached themselves to the *apparently* 'simple' bodies (Fire, Air, Water, and Earth). (*De generatione et corruptione*, Book II, 3.)

Aristotle created this model so that he could explain more easily how some simple physical changes, like boiling and condensing, come about. In this way his theory was more scientific than Plato's because

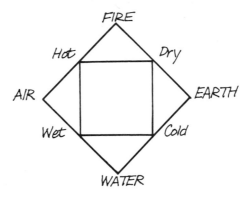

4 simple bodies (or elements) on the large square, 4 qualities on the small square

it was designed to explain what we can actually sense and see happening. However, he was careful, in the paragraph quoted above, to speak of only *apparently* simple bodies and in the next paragraph he goes on to explain that the pure element does not exist in the perceptible world. What we sense as 'fire' in nature is a mixture, the complete purity of 'fire-ness' could only be found in the simple body. This, one is left to suppose, could again only be perceived through the 'eye of the soul'!

The Greek Legacy
The works of Plato and Aristotle were revered for nearly two thousand years—Plato as a philosopher and Aristotle as a logician and scientist. But Aristotle had studied under Plato for twenty years and though he criticised some of his master's views he never completely threw off his influence. How far he was from being a scientist in the modern sense can be seen very clearly in the following paragraph:

Obviously it is the province of a speculative science to discover whether a thing is eternal and immutable and separable from matter; not, however, of physics . . . nor of mathematics, but of a science prior to both. Hence there will be three speculative philosophies: mathematics, physics and theology . . . the most honourable science must deal with the most honourable class of subject. (*Metaphysics*, VI, i.)

When Aristotle died in 322 BC he left a flourishing academy, the Lyceum, in Athens. Other philosophers followed him and there was considerable scientific activity there for nearly a century more but little was achieved that was handed on to subsequent generations, since a rival centre of learning was emerging in Alexandria which was to have a far greater influence. Before he founded the Lyceum, Aristotle himself had been tutor to the young Alexander of Macedon who was later to conquer almost all the known world and in whose name the new city of Alexandria was founded. Here arose a new centre of learning—the Museum. It was a much bigger and richer academy than the Lyceum had ever been and included a library of, perhaps, half a million scrolls of papyrus. These included Aristotle's personal collection and thousands of works long since destroyed or lost. No wonder it was considered one of the wonders of the ancient world! Here a new phase in the study of matter began which owed much to earlier Greek thought.

Before we study the new growth of science it is worth enumerating the outstanding ideas that the Greek passed on to later scientists in this field.

1 There is a basic *immutable substance* that underlies all matter which is never destroyed but merely changes its form. This gave rise to the unstated principle of Conservation of Matter.

2 The visible substances of the world all derive from the *four elements*. An element was thought of as a simple, unmixed substance from which other more complex substances arise by combination or mixing.

3 Matter is composed of tiny 'uncuttable' atoms. Sometimes this was included with the four-element theory so that it was the atoms

of these elements which intermingled. The atomic theory had several distinguished supporters but was never very widely accepted.

These three concepts were to prove immensely valuable later when they had been defined more explicitly and were tested experimentally. However, in addition to these, there were two other, less useful, concepts that carried considerable weight.

4 The *four qualities* of matter (hot, cold, moist, and dry) could be interchanged thereby changing the nature of the substance to quite another substance.

5 The *realities of the world* are more comprehensible to the *mind* than to experiment or to the senses. This was the legacy that Alexandrian science was to inherit from the Greeks in the field of chemistry.

3
The Birth of Alchemy

Alexandria was a 'melting-pot' of cultures, like America has been in the modern world. Apart from the native Egyptians there were Greeks, Jews, and Romans all living there as well as travellers along the trade routes from farther east. Each group influenced the general climate of thought and the study of matter that emerged—named *Alchemy* after the Land of Chem (Egypt)—was very different from what had preceded it.

The Egyptians had a long tradition of working in metals. In direct contrast to the Greeks who relegated the practical crafts to slaves or social outcasts, Egyptian metalwork and medicine were closely associated with the priestly caste. Formulas for the purification of metals were said to be inscribed in hieroglyphics on the walls of the tombs of the Pharaohs. Papyri containing medical recipes were found in a tomb in the sacred city of Thebes dating back to about 1500 BC, probably the oldest 'book' in the world! Since another priestly craft was astronomy it is not surprising that the seven metals best known in the Ancient World, gold, silver, iron, mercury, lead, tin, and copper came to be associated in an astrological sense with the Sun, the Moon, and the five planets visible to the naked eye. Another local skill was that of dyeing, not only of materials but also of wax for use in painting, and of the metals themselves.

Alongside the religious slant in the crafts were various alien religions which came to have a great influence on the way in which alchemy was practised. Greek official religion had become, by Plato's time, so mixed with discreditable tales of the doings of the 'Olympian' gods that it was difficult to believe in it deeply in any personal sense, or even to detect in it a pattern of behaviour that was 'pleasing

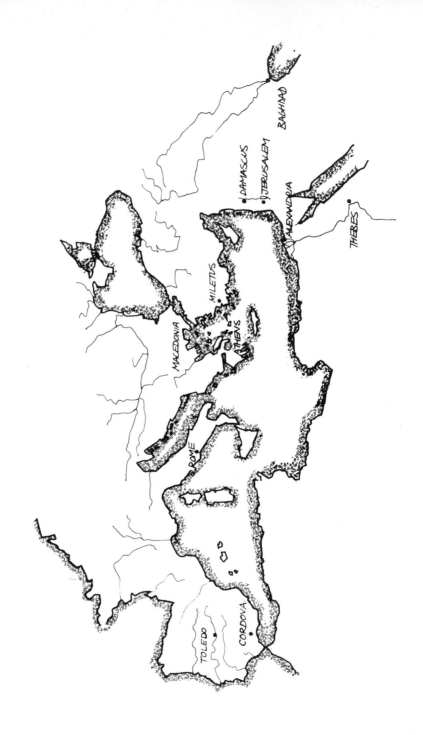

to the gods' as Plato's Socrates pointed out. In Alexandria three quite different religions had an important following, the cult of Mithras, Judaism, and Christianity. These were at the same time mysteries of an omnipotent, unknowable God and also codes of personal conduct. Religion became both ecstatic and intimate. The influence of science on religion can be traced in the Book of Wisdom in the Apocrypha. It is wrongly attributed to King Solomon and is thought to have been written by an Alexandrian Jew or Jews about 50 BC.

> For earthly things were turned into watery, and the things, that before swam in the water, now went upon the ground.
> The fire had power in the water, forgetting his own virtue: and the water forgat his own quenching nature.
> On the other side, the flames wasted not the flesh of the corruptible living things, though they walked therein; neither melted they the icy kind of heavenly meat, that was of nature apt to melt.

Wisdom may be the 'hand-maiden of the Lord' but it is also held to include a knowledge of the Greek four-element theory! Below is a fragment of a prayer to Mithras probably in vogue in Alexandria about AD 200.

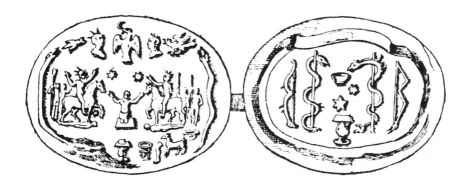

Mithraic Cameo

First Beginning of my beginning, First Principle of my principle; Breath of breath, First Breath of the breath within me; Fire, which among the compounds which form me was given by God for my own compound, First Fire of the fire within me; Water of water, First Water of the water within me; Earthly Substance. . . .

This prayer gives a very good foretaste of how science was to combine with religion during the long reign of alchemy in a curious combination of personal ecstasy and rudimentary chemistry that is very hard for us to tolerate today.

The earliest alchemists saw in Plato's doctrine of resemblance to eternal forms a theoretical basis for their belief that doing anything which gives a metal the quality of 'gold-ness' is, in fact, changing it to gold. Since the 'ideal' gold cannot be recognised by the senses, they believed that the *appearance* of gold is all that can be striven for or achieved. In the earliest manuscripts extant they seem clear that the changing or transmuting of a metal into gold was very like a dyeing operation.

. . . Having obtained the metal [copper] of clear tint, yellow it with anything until a yellow [gold] colour is obtained. Cast this upon any kind of metal, for copper of clear tint on becoming yellow tinctures every kind of metal [to a golden colour].

This recipe is attributed to one Bolos Democritus, a Greek living in Egypt in, perhaps, the first century AD.

The Ecstasy begins

A religious extension of Plato's theory was then made where the 'idea' of a substance which can be perceived only by the 'eye of the soul' is itself the spirit or soul of the substance. Indeed we still use terms like 'spirituous liquid' and 'spirits of salt' (hydrochloric acid) which are directly traceable to these early times.

'Mercury robs all metals of their appearances. Just as wax takes the colour which it has received, so mercury whitens all metals and attracts their souls.'

26

This fragment is based on good observation. It is a fact well known and irritating to any woman working in a laboratory that mercury actually dissolves into other metals (eg, a gold ring) and gives them a dull flat silver colour. The alchemists, however, saw this in terms of conflict and soul! The passage was written in the second or third century AD and the tendency to write of alchemy in allegory, dream-imagery, and anthropomorphic terms rapidly became more pronounced. While some practical chemical work was probably still going on, the mystical element in alchemy ran riot! The following account is, at first sight, merely a particularly horrible nightmare; in fact it is intended to be read as a chemical account of the oxidation of a metal to a powder followed by heating with charcoal to regain the original metal!

> I fell asleep and saw before me a priest (who said) . . . 'I am Ion, priest of the sanctuary, and I have suffered intolerable violence. For one came quickly in the morning, cleaving me with a sword, and dismembering me systematically. He removed all skin from my head with the sword that he held; he mixed my bones with my flesh and burned them with the fire of treatment. It is thus, by the transformation of the body, that I have learned to become spirit'.

It remains only to describe some of the positive, practical achievements of this early period of alchemy. Methods of purifying metals were examined together with methods of dyeing and alloying them. The alchemists studied the effects of sulphur, in the vapour as well as 'sulphur water'. They had formed hydrogen sulphide and reported its nauseating smell and had heated metals in an atmosphere of mercury vapour. For these experiments they had, no doubt, developed furnaces whose heat could be carefully controlled. There were also many recipes that required sealed containers, indeed these formed so large a part of the slow heating processes advocated by the alchemists in the name of their legendary founder, Hermes Trismegistos, that we still use their phrase 'hermetically sealed' to this day.

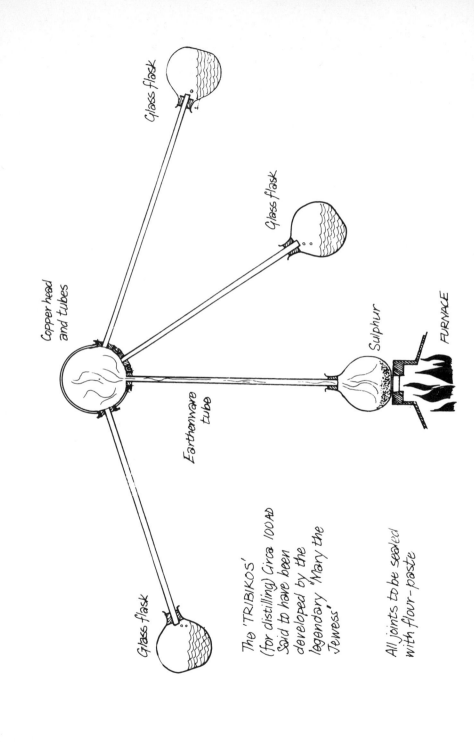

Glass flask

Glass flask

Copper head
and tubes

Sulphur

FURNACE

Earthenware
tube

Glass flask

Glass flask

The 'TRIBIKOS'
(for distilling) Circa 100 AD
Said to have been
developed by the
legendary "Mary the
Jewess."

All joints to be sealed
with flour-paste

Looking back from modern times to those early days of chemical experiment it is easy to point out their lack of direction. After the hypotheses of Empedocles the essential job clearly was to test his four-element theory in practice. It had not been founded on experimental evidence and this is what the alchemists should have set out to do. Instead of accepting his theory they should have analysed the materials around them, it seems to us, to see if they really were made up of four simple elements. It is too easy to be wise so many centuries later! The world contains many complex and impure substances; the Alexandrian chemist had little apparatus and less knowledge; it was inevitable that progress should have been very slow indeed.

The influence of religion and mystery made progress even slower. The alchemists' reports of their experiments became so loaded with dream symbolism that, in the twentieth century, Dr Jung found striking similarities between them and the dreams of his patients. However fascinating this thought may be, it was hardly conducive to clear thinking and scientific reporting! The main drawback of the pseudo-religious attitude of the alchemists was that, if the position of the planets, the music being played, and the purity of the experimenter, could all influence the outcome it would be hard to pin down the chemical factors—which had, in fact, actually caused the reaction to take place. Any experiment, to be of value, must be repeatable, by anyone under specified laboratory conditions. This point, obvious to us, was not clear to the alchemists. Their experiments were designed largely in the hope that suitable symbolic processes would cause gold to 'grow' out of other metals in a way parallel to the growth of man's spirit.

4
Alchemy in the Arab World

The year AD 610 was to be of great significance to the whole Middle East. This was the year in which the prophet Mohammed began his religious activities in Arabia. Soon Mecca was to fall to him and Islam became a force which united the Arabs and led, within a century, to the formation of an empire stretching from Persia to Spain. In AD 640 Alexandria, the centre for science, fell to the Arabs without a gesture of help from Byzantium under whose rule it had been for several centuries. The great library of the Museum, which had already been ravaged more than once, was finally destroyed. It is said that the manuscripts in it were used to heat the public baths for several months! This loss was, and still is, a terrible blow to our study of early philosophy but within a few years new academies of learning were set up in Persia. Greek works still extant were translated into Arabic by Nestorian Christians living in Syria and there was a slow trickle of Greek books imported from Byzantium.

The earliest Arabic alchemist of whom we know anything was a Prince Kalid who lived in Damascus. Legend has it that, after years of fruitless endeavour, he witnessed the transmutation of base metal into gold and learned the secrets of alchemy from an Alexandrian monk who was attempting to convert him to Christianity. This prince lived about AD 660–704 while the work of translation was still in progress. The flowering of Arabic alchemy was to come a century later.

The name of the Caliph Haroun Al-Rashid is known in the West from the stories of the 'Thousand and one nights'. It was at his court that one of the greatest of the Arabian alchemists flourished—Jabir, who was also known in Latin translation as Geber. He was court physician in Baghdad about AD 770–803 and is credited with an enormous output of work on both alchemy and medicine. Jabir seems to have been a member of a mystic religious sect in Islam which was much influenced by neo-Platonic thought. Following Plato's ideas of 'soul' in an ecstatic religious environment may seem a very poor educational background for a scientist; but, curiously enough, Jabir seems to have had a natural instinct for experiment. Indeed he instructed his followers that it was the first duty of an alchemist to test his theory in practice. He sets out some very clear instructions for the preparation of many new compounds (nitric acid, sal ammoniac, and acetic acid). Here, for the first time, we meet recipes with clearly indicated weights of substances to be added. This could have been the beginning of real chemistry but the flavour of the times was still too mystical and credulous for this aspect of Jabir's work to take root.

The Sulphur Mercury Theory and the Philosopher's Stone

In alchemical circles Jabir's lasting fame lay in his doctrine of the 'sulphur-mercury' composition of metals. The effects of sulphur and mercury on metals had been closely studied in Alexandria and the

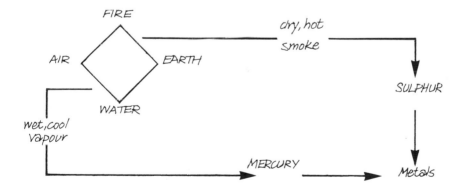

glorious sequence of colours obtained by heating these two substances together had probably been known for some time.

The dry yellow powder, sulphur, that burns away to acrid fumes must have seemed the complete opposite (in its Aristotelian qualities) to the strange, cool, liquid metal, mercury, which colours other metals but never itself catches fire. Jabir accordingly formed the theory that, since sulphur was *hot* and *dry* and mercury was *cool* and *wet*, the combination of these two was enough to give all the qualities of metals. The only problem was—what are the correct proportions in which they should be combined to form a particular metal? Jabir attempted to solve this puzzle with the aid of a magic Number Square and a key of numbers attached to each letter of the Arabic alphabet.

All rows and diagonals add up to 15

Then, by spelling out the name of the metal required, he could refer to the Magic Square to find out the appropriate amounts of hotness, wetness, dryness, and coldness required! Alternatively it should be possible, by altering the proportions of sulphur and mercury, to change one metal into another. Jabir taught that this process was to be brought about by a suitable 'elixir' (an Arabic word we still use). This seems to be the starting-point for the centuries-long hunt for the mythical 'Philosopher's Stone' that would transmute base metals into gold.

Persian Peak of Scholarship

Since alchemy was practised in Persia by eminent medical men and their pupils it acquired a reputation and erudition that it had not enjoyed before. Among other theories the perennial Atomic theory of Democritus was read and reappraised by a famous physician of the ninth century—Razi. He taught that the atoms of the four elements move about in the empty spaces that exist between them and that all the properties of hotness, coldness, hardness, colour, etc, arise from the effects of these motions. There was little new in Razi's teaching but it serves to show how the old speculation was kept alive through the long centuries between the ancient Greek philosophers and the Scientific Renaissance in Western Europe. The

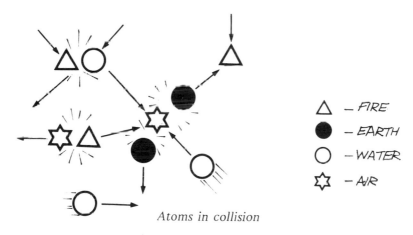

△ — FIRE

● — EARTH

○ — WATER

☆ — AIR

Atoms in collision

climate of thought in Baghdad must have been energetic and liberal. Differences of opinion were tolerated, sceptics published their doubts and the orthodox replied with their views, and through it all chemical apparatus grew more abundant and experiments more sophisticated.

By the end of the tenth century the Arabic empire was showing signs of strain, but scholarship was still flourishing. The great Persian philosopher, Avicenna, wrote and taught at this time on medicine, logic, mathematics, physics, music, and alchemy. Although he endorsed Jabir's mercury-sulphur theory of metals he was an avowed sceptic as far as transmutation was concerned. He insisted that, though the colour of a metal could be altered, its essential nature remains unchanged. Nevertheless, in spite of the fame attaching to Avicenna's name, transmutation continued to be fervently sought and wistfully hoped for by the majority for another six centuries; and indeed up to the present day, it is still possible to find the dark laboratories of hopeful alchemists in Damascus and Baghdad!

Into darkest Europe
Modern science is a proud achievement of Western Europe, but in the long span of time from about AD 300 to 1200 the search for knowledge about the workings of the universe was dormant there. Technical progress took place and the crafts flourished, but of scientific speculation, hypothesis, and experiment we have no record. However, in the tenth century the hub of Arabic learning moved into Spain. In Cordova and Toledo, in particular, universities were set up which gradually attracted scholars from all over Europe. First there was a flowering of interest in Arabic alchemy, medicine, and logic amid a cultured, tolerant environment in which adherents of all the three great religions participated. Then more fanatic Moslems from the Sahara drove the 'infidel' Jews and Christians farther north. Toledo was recaptured from the Arabs and here, in a predominantly Arabic-speaking region, Christian scholars of the twelfth century, from as far away as Greece and Yugoslavia as well as from France and England, eagerly rediscovered the scientific works of the Arabs and the Greeks.

The task was not easy. Few scholars knew both Arabic and Latin and, in some cases, the manuscripts had first to be translated into Hebrew by the Spanish Jews and then into Latin. Books on alchemy were the first to be translated but Europe was most eager to read the works of Aristotle. Only one book of his, on Logic, had survived the Dark Ages. Unfortunately the works of the ancient Greeks that the Arabs possessed were, by this time, so buried beneath later additions and commentaries that the translations produced would have come, in some cases, as a great surprise to the original authors!

Soon after the scholars of Toledo finished their task of translation more reliable manuscripts were being obtained from another source. In the fourth crusade Christian Europe had recaptured Constantinople (AD 1204). By the end of the century translations from Greek into Latin were available which were recognised as more reliable and were eagerly read.

The first European university to be set up which studied anew the scientific works of the Greeks was at Paris. Here Thomas Aquinas taught his new synthesis of Aristotle and Christian theology and the Englishman Roger Bacon lectured on the duty of purifying Christendom by learning in order to confound the pagans! By 1255 the *Physics* and *Metaphysics* of Aristotle had become required reading for all the students of the University of Paris. The new university at Oxford rapidly followed this example, although reliable texts were still hard to come by.

> If I had power over the books of Aristotle as at present translated, I would burn them all; for to study therein is but lost time, and a source of error and a multiplication of ignorance beyond all human power to describe. And, seeing the labours of Aristotle are the foundation of all wisdom. . . . Whosoever will glory in Aristotle's science, he must needs learn it in its own native tongue, since false translations are everywhere, in theology as well as in philosophy. (Roger Bacon—*Compendium Studii Philosophiae*, 1271.)

Roger Bacon was an exception as a scholar for his interest in alchemy and in experiments, although it is difficult to assess how much practical work he actually did.

The lure of alchemy worked potently in Europe as Chaucer bears witness in the *Canterbury Tales* but little new was added to the theory. It was only to be expected that, for at least two centuries after the introduction of the 'Divine Art' into Europe, budding alchemists were content to study the old theory and perform experiments along the lines that Jabir and others had laid down. The mysticism, secrecy, and symbolism that had so dogged alchemy since

Breath from heaven making the Philosopher's Stone enters the Hermetic Vase containing sulphur and mercury. The seven metals drip to earth and are received by the spirit, soul and body of Man.
From an alchemical illustration, Oxford 1652

its beginnings in Alexandria persisted among the more esoteric of the brethren while, lower on the social scale, 'puffers' worked at their laboratory furnaces trying to effect transmutations. Some level-headed citizens were sceptical and some alchemists disillusioned, but for several centuries no scientific speculation appeared to challenge the theory behind the search for the Philosopher's Stone.

'Tel me the name of the privy stoon?'
And Plato answerde unto him anoon,
'Tak the stoon that Titanos men name'

'It is a water that is maad, I seye,
Of elements foure' quod Plato.
'Tel me the rote, good sir: quod he tho,
'Of that water if that it be your wille?'

Thanne conclude I thus; sith god of Hevene
Ne wol nat that the philosophres nevere
How that a man shal come un-to this stoon,
I rede, as for the beste, lete it goon.

(The Canon Yeoman's Tale, *Canterbury Tales*, Chaucer.)

5
The Renaissance of the Atom

While the old books were trickling into Europe the attention of potential scientists seemed to be almost completely absorbed in trying to understand and assimilate the old ideas. No new theories of matter appeared. During the next three centuries Aristotle's influence in particular was so great that his conclusions were rarely, if ever, challenged. Plato's works were next to be rediscovered. Following his theories on education, tutors, schools, and colleges gave more attention to the study of mathematics than ever before. In the fifteenth century the long poem of Lucretius, *De Rerum Natura*, was printed. He was an ardent follower of the old Greek atomists and there was an immediate revival of interest in the Atomic Theory.

> *. . . Do but observe:*
> *Whenever beams make their way in and pour*
> *The sunlight through the dark rooms of a house,*
> *You will see many tiny bodies mingling*
> *In many ways within these beams of light*
> *All through the empty space, and as it were*
> *In never-ending conflict waging war,*
> *Combating and contending troop with troop*
> *Without pause, kept in motion by perpetual*
> *Meetings and separations; so that this*
> *May help you to imagine what it means*
> *That the primordeal particles of things*
> *Are always tossing about in the great void.*
> (De Rerum Natura c 60 BC.)

Archimedes and division into Grains
The sixteenth century saw the beginning of wide-scale distribution

of knowledge by printed books. Among those printed were new translations of the works of Archimedes which were to have an enormous influence. Greek mathematics had been very successful and highly esteemed by philosophers of all ages and Archimedes was, undoubtedly, one of the most brilliant and original of those mathematicians. In addition to his practical inventions—'marvels', of which there are many legends—his work shows how far he had gone towards casting off the traditional Greek disapproval of the 'infinitely great' and the 'infinitesimally small' and was anticipating, to some extent, the discovery of Calculus which was to be the crowning achievement of seventeenth-century mathematics. Archimedes invented a new system of 'orders of magnitude' to cope with very large numbers and wrote a short book using it. The intriguing title of this work is *The Sand Reckoner*. It begins as follows:

> There are some, King Gelon, who think that the number of the sand is infinite in multitude; and I mean by the sand not only that which exists about Syracuse and the rest of Sicily but also that which is found in every region whether inhabited or uninhabited. Again there are some who, without regarding it as infinite, yet think that no number has been named which is great enough to exceed its multitude. And it is clear that they who hold this view, if they imagined a mass made up of sand in other respects as large as the mass of the earth, including in it all the seas and hollows of the earth filled up to a height equal to the height of the highest mountains, would be many times further still from recognising that any number could be expressed which exceeded the multitude of the sand so taken. But I will try to show you by means of geometrical proofs, which you will be able to follow, that, of the numbers named by me and given in the work which I sent to Zeuxippus, some exceed not only the number of the mass of sand equal in magnitude to the earth filled up in the way described, but also that of a mass equal in magnitude to the universe.

He concludes his calculation, 'Hence the number of grains of sand which could be contained in a sphere of the size of our universe

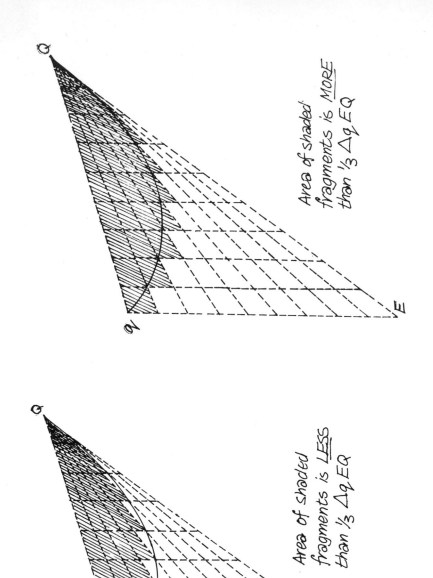

Area of shaded fragments is LESS than $\frac{1}{3} \Delta q\, EQ$

Area of shaded fragments is MORE than $\frac{1}{3} \Delta q\, EQ$

is less than 1,000 units of the seventh order of numbers.' (10^{51} as we should now write it, that is 51 noughts after the unit one.)

More important than this, is the way in which Archimedes used the method of dividing finite areas into smaller and smaller pieces in order to simplify the shape involved and find its area by adding up the area of the pieces. Division of a finite quantity into a potentially infinite number of infinitesimally small ones is the essence of integration in Calculus without which Newton's Theory of Gravitation would never have been formulated. With reference to figure on p 40:

To find the area of the sector of a parabola qEQ

(Archimedes)

Assumption, implicit but not stated for fear of offending against the mathematical proprieties of the times, that if there were a much greater number of much smaller pieces they would correspond more closely to the sector. Therefore the sum of an infinity of infinitesimal shapes would, in the limit, fit the sector exactly.

Area of the sector of a parabola $= \frac{1}{3} \triangle$ qEQ.

Galileo

At last the time was ripe for the rebirth of European science. The cultural Renaissance had flowered and passed without any corresponding originality in scientific speculation. It is true that the Polish monk Copernicus had stood the old astronomical theories on their heads by his hypothesis of a sun-centred universe but this was almost the only challenge that there had been to ancient Greek science. In the sixteenth century there arose in Italy a one-man revolution. Here was a mind brilliant, inventive, and bold which seized upon the most fertile ideas of the old world, freely added his own and so launched the Scientific Renaissance. This man was Galileo Galilei.

Galileo followed Archimedes in many ways and with avowed admiration—'O superhuman Archimedes!' The Atomic Theory of the

Greeks also appealed to his mechanical mind. He fully accepted Democritus' dictum: 'By convention colour, by convention sweet, by convention bitter, in reality nothing but atoms and the void.' Although he also accepted Democritus' idea of 'subtil fire atoms' he

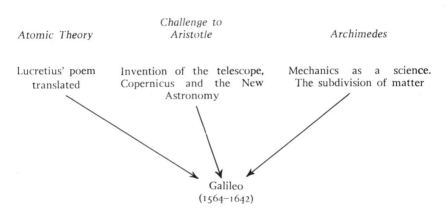

made a great step forward by asserting that the action of heat is accompanied by movement which is transmitted to the atoms.

Salviati: Gold and silver . . . do not become fluid until the finest particles of fire or of the rays of the sun dissolve them, as I think, into their ultimate, indivisible, and infinitely small components. . . .

Sagredo: . . . with regard to the surprising effect of solar rays in melting metals, must we believe that such a furious action is devoid of motion, or that it is accompanied by the most rapid of motions?

Salviati: We observe that other combustions and resolutions are accompanied by motion, and that, the most rapid; note the action of lightning and of powder used in mines and petards; note also how the charcoal flame, mixed as it is by heavy and impure vapours, increases its power to liquefy metals whenever quickened by a pair of bellows. Hence

42

I do not understand how the action of light, although very pure, can be devoid of motion.

(*Dialogues Concerning Two New Sciences*, 1638)

The problem of the 'void' or vacuum between the atoms continued to be a difficult point in the theory as it had been to the Greeks. In Galileo's time the barometer was not understood and the air pump had still to be invented. He stood by the medieval principle that 'Nature abhors a vacuum' and used it to explain the existence of the cohesive forces between the atoms in a solid. Just as two smooth marble slabs are hard to separate because this would cause a momentary vacuum between them, so, he thought, the tiny spaces between the atoms which are quite empty make solids strong and difficult to break.

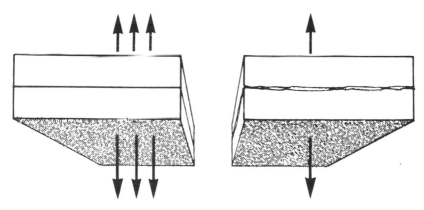

The smooth marble slabs on the left are hard to separate ('nature abhors a vacuum'), while the rough slabs on the right with air between can be separated easily

The next generation of physicists was to learn much more about forces and about the nature of a vacuum, but Galileo's theory was an advance in that it attempted to explain the properties of matter mechanically. The Atomic Theory had become the accepted hypothesis of the nature of matter by the end of the seventeenth century which was long before there was any real experimental evidence for the existence of individual atoms.

Now, at last, the hoary doctrines of alchemy could be challenged. Atoms were tiny and restless but *permanent* particles; there was no room for the changing 'qualities' of substances in this theory. Contemporary alchemy had concentrated on the sulphur-mercury synthesis of metals, to which salt had been added to match the trinity of spirit, soul, and body of man. The new generation of scientists were frankly unimpressed by Aristotle's reputation and saw any mixture of substances as a simple mingling of distinct and separate kinds of stone. Galileo had written:

> I never was thoroughly satisfied about this substantial transmutation whereby matter becomes so transformed that it is necessarily said to be destroyed, so that nothing remains of its first being, and another body, quite differing from it should then be produced. If I fancy to myself a body under one aspect, and by and by under another very different one, I cannot think it impossible that it may happen by a simple transposition of parts without corrupting or engendering anything new.

6
The English Contribution

By the middle of the seventeenth century the main scene of scientific discovery had shifted north into England. Here was a country seething with religious discord, civil war, widespread superstition, and a daily proliferation of political pamphlets. All men, academics included, gave easy credence to the most unlikely stories. In 1629 no less a person than the Vice-Chancellor of Cambridge University examined publicly, as a serious portent of terrible disaster, the case of a book of religious mysteries which had been found, it was said, in the belly of a cod-fish! England was gripped by an intellectual conservatism deeper, in many ways, than on the continent of Europe. The works of Aristotle still reigned supreme in most university circles and the news of Galileo's discoveries made little impact on the general climate of thought. Belief in alchemy and astrology was almost universal, as was the endless search for perpetual motion. The subject of research that raised the most academic enthusiasm was still the tracing out of biblical prophecy, from the Book of Revelation and elsewhere! Yet it was out of this untidy welter of speculation that a most exciting period of scientific progress arose and if we pick out of this profusion only those ideas which time has proved to be of real value it takes on the air of a 'Golden Age' for science. Reading the lives of great men of the times proves, however, that there was no sudden scientific revolution. The same men who inaugurated the brilliant 'New Philosophy', men of the calibre of Isaac Newton and Robert Boyle, indulged in a pursuit of the ancient mysteries with almost equal enthusiasm.

A specifically English line of inquiry was emerging which was neither academic nor theoretical—the seventeenth-century men of science were thrilled by *practical* experiments. France, under the leadership of the great Descartes, was now the centre of philosophy but even their 'Natural Philosophy' smacked more of Platonic theory about how the intangible universal element aether must be distributed than of a true description of Nature. In England the century had begun with a practical work about magnetism by Queen Elizabeth's personal physician, Dr William Gilbert. This was followed by the publication of a new system of scientific philosophy by Francis Bacon putting forward a plea for the practical mastery of science. It is well known that this philosopher met his death as a result of a chill caught while stuffing snow into a dead chicken with his own hands in order to find out if refrigeration delays putrefaction! This readiness to carry out tests and develop practical skills became a characteristic of the age. In 1662 when the Royal Society was only two years old a curator was engaged to entertain the noble members each week 'with three or four considerable experiments, expecting no recompense till the Society get a stock enabling them to give it'.

'The Sceptical Chymist'

One of the founder-members of the Royal Society was Robert Boyle, a life-long patron of science, a man of great personal piety and one in whom the curious contradictions of the century were most apparent. He was a wealthy man who built himself an excellent laboratory, employed several assistants and was an ardent and careful experimenter with little respect for the practical results of the alchemists. In 1661 Boyle, at the age of thirty-five, wrote a scientific classic attacking alchemy called *The Sceptical Chymist*. It is a dialogue between a character holding traditional views, who is allowed to ask a few questions, and the much more voluble mouthpiece of the author! The work is polemic and realistic, if rather longwinded. Early in the book Boyle attacks the alchemical idea of transmutation by heating with characteristic vigour:

46

I do not mean that anything is separable from a body by fire that was not materially pre-existent in it; for it far exceeds the power of meerly naturall agents, and consequently fire, to produce anew, so much as one atome of matter, which they can modifie and alter, not create. . . .

Alchemy died slowly; it had survived so many centuries since its birth in old Alexandria, but from now on science was awake and ready to search continuously for new and better theories of the elements of matter.

Boyle launched the search by providing a new, clear definition of the term 'element', a word which had been freely bandied about by the alchemists as they echoed the early guesses of Empedocles on the ingredients of matter. True to his vaunted 'scepticism' Boyle is careful to name no particular substance as an element, and rightly so, for chemical experiment was still in its infancy and the evidence that was needed was not yet to hand.

I mean by elements . . . certain primitive and simple or perfectly unmingled bodies; which not being made of any other bodies, or of one another, are the ingredients of which all those called perfectly mixt bodies are immediately compounded, and into which they are ultimately resolved; now whether there be any one such body to be constantly met with in all, and each, of those that are said to be elemented bodies, is the thing I now question.

His own experiments did not support the idea that there were either three, four, or five elements. The situation was very complex and he could only state his own opinion that there might well be many different elements making up all the varied compounds then known.

In later years Boyle lost the clear-sighted sceptisim of his youth, he even ceased to value it. Others were won over by his book but he grew steadily more conservative with age until he reverted in the end, to the prevailing mystery of alchemy. Shortly before his death he wrote to a friend:

... I had a kind of ambition (which I now perceive to have been a vanity) of being able to say that I cultivated chemistry with a disinterested mind, neither seeking nor scarce caring for any other advantages by it, than those of the improvement of my knowledge of nature. . . . But, however, since I find myself now grown old, I think it time to comply with my former intention to leave a kind of Hermetic legacy . . . more of kin to the noblest Hermetic secrets. . . .

When he died Boyle bequeathed to Newton a sample of his precious 'red earth' which, when mixed with mercury, would, he believed, turn into gold! In this dynamic century it was only the forceful and obstinate young who could stand out against the appeal of traditional, medieval thought. Isaac Newton was to show exactly the same sad reversion to the ecstatic studies of earlier times as he grew old. But if Boyle and Newton both repudiated the challenging attitudes which they had adopted as young men they never wavered in their respect for experimental results and each man had many of these to his credit.

The Void between the Atoms

Between 1643 and 1660 many experiments had been performed with barometers and air pumps to show that a vacuum really can exist. Torricelli was the first to explain the simple mercury barometer. Pascal confirmed that it was the pressure of air that held up the column of mercury by taking a barometer up a mountain where less mercury was held up by the more rarefied air.

Von Guericke invented the first air pump and, with superb showmanship, demonstrated in public that it took sixteen horses to separate the two halves of an evacuated sphere!

Robert Boyle himself was very enthusiastic about the air pump, using a barometer to record the pressure. He examined breathing, burning, and falling in a vacuum as well as the expansion and contraction of air. These experiments not only led him to formulate the famous (or 'infamous' to so many schoolchildren!) law which bears his name,* but to appreciate the real 'springiness' of air.

If anyone doubts that 'thin air' is elastic they need only compress some air in a simple bicycle pump (placing a finger over the open end) and then let go! Boyle would have liked to have been able to explain this by his favourite 'corpuscular theory', but the motion of atoms in a gas was not yet understood. He was forced to picture the particles of the air as tiny coiled springs.

However, Boyle's work was very positive and far-reaching although it was not widely accepted until the usual 'digestion period' had taken place. He had cast a lot of doubt on the very basis of alchemy, defined the chemical term 'element' and, most important of all, had demonstrated that a vacuum, identical to the supposed void between the atoms, can be shown to exist. This had always been one of the 'sticky' points in the Atomic Theory from Democritus to Galileo. The idea of complete emptiness seemed to induce a kind of mental vertigo in philosophers throughout the centuries! It was they, not Nature, that 'abhorred a vacuum' and Boyle's systematic series of experiments were necessary to clear the way for the resurgence of the Atomic Theory.

Isaac Newton

The man who overshadowed the seventeenth century, casting his influence across the next two hundred years of science, was Isaac Newton, a figure so austerely intellectual and brilliantly novel in his work as to stir wonder and even reverence to this day. He left behind diaries, notebooks, and unfinished manuscripts in great quantities and from them we can attempt to understand this man, intellectually a giant and yet with a character strangely insecure and withdrawn. He was born posthumously in 1642 (the same year as Galileo died) so weakly and premature a baby that the midwife almost despaired of his life. A modern psychologist might note with

---

* 'The volume of a given mass of gas kept at constant temperature varies inversely as the pressure.'

49

Sir Isaac Newton painted by Sir James Thornhill

interest that within two years his mother left him in the care of his grandmother when she married again and went to live with her new husband a mile distant. He was known to have been extremely fond of his mother and it would be easy to suppose that this early deprivation of her presence sowed the seeds of a life-long habit of reserve which made him shun both affection and friendship. Such 'potted psychology' is probably very suspect; nevertheless a picture does emerge of Newton both as boy and man of such austere piety and such morbid fear of criticism that his only abiding pleasure was in private meditation. Perhaps it was fortunate for humanity, if not for him, that there were no contemporary psychoanalysts to expose and heal these defects of character, for it may have been by 'repressing' these normal channels of sociability that he gained some of his enormous power of mental concentration. So many great men have had these abnormal psychological traits that it seems almost a law of nature that only by sealing up the usual outlets of emotion can the occasional man generate within himself this powerhouse of creative energy.

As a child Newton disliked games and made few friends, in his late teens he was engaged to a childhood friend but he never married. Years later when overwork brought on a nervous breakdown he suffered from delusions of persecution and wrote this pathetically revealing letter to the philosopher Locke:

> Sir,
>   Being of opinion that you endeavoured to embroil me with women and by other means, I was so affected with it, as that when one told me you were sickly and would not live, I answered, 'twere better if you were dead'. . . .

He recovered from this illness but throughout his life he sought little contact with people but only with ideas. Though many of the Cambridge legends of his absent-mindedness may be exaggerated we do know that he was often so bewitched by a train of thought that he forgot to sleep or eat and that his dress was notoriously untidy.

Yet it was to this superb quality of single-minded absorption that he himself referred when asked to what he attributed his great achievements: 'If I have any genius not common to other men it lies in the fact that when an idea first comes to me, I ponder over it incessantly until its final results become apparent.'

Newton's life shows the same decline from a youthful pinnacle of creative thought that is the lot of most mathematicians. From 1665 to 1667 the University of Cambridge closed down due to the ravages of the plague and Newton returned home to the peace and privacy of his manor-house at Woolsthorpe in Lincolnshire. Here during two 'golden years' at 23 years of age he invented his method of 'fluxions', the first formulation of mathematical calculus, he used it to design the laws of motion and his great theory of Universal Gravitation, and he carried out almost all his new work in the field of optics. None of this was published until many years later. It took a combination of the pleading of his friend Edmund Halley and the sting of another's related discovery to force him into setting down his results in the famous *Principia* and then, perhaps by deliberate intent, it was found to be so difficult that almost none of his contemporaries could follow it. The king is reported to have remarked that 'like the Peace of God it passeth all understanding!' Newton seems to have taken little pride in his great achievement and reacted so violently to criticism about his *Opticks* that he wished to give up science altogether. He wrote: 'I have long since determined to concern myself no further about the promotion of philosophy' and 'I see a man must either resolve to put out nothing new or to become a slave to defend it.'

In fact during his long life he wrote far more about theology than about physics and probably placed a greater value on his analysis of revelation in the Book of Daniel and on the scientific theories of his youth.

Newton was a kind and generous man, supporting many nieces and nephews but his chief pleasure, in common with most English scientists of his time, was in practical research. As a child he was remembered for the model clocks that he made, he always ground his own lenses and it was on the strength of the new telescope that he both designed and constructed that he was elected to the Royal Society. His accounts of experiments on light and colour are a joy to read but his chief pastime throughout his life was experimenting in chemistry or alchemy. Here he could find a strange combination of scientific research and occult mystery which seemed to knit together his greatest passions. His assistant wrote of him

> . . . especially at spring and fall of the leaf . . . he used to employ about six weeks in his elaboratory, the fire scarcely going out either night or day; he sitting up one night and I another, till he had finished his chemical experiments, in the performances of which he was the most accurate, strict, exact. What his aim might be I was not able to penetrate into, but his pains, his diligence at these set times made me think he aimed at something beyond the reach of human art and industry.

What the assistant vaguely guessed must always have been true of Newton as of all the scientists of his generation. Scientific research was the pursuit of Truth and so of God Himself—a Christian duty as well as an intellectual pleasure.

> For so far as we can know by natural Philosophy what is the first Cause, what Power he has over us, and what Benefits we receive from him, so far our Duty towards him, as well as towards one another, will appear to us by the Light of Nature. (Newton, *Opticks*.)

Forces between the atoms

Isaac Newton's greatest contributions to science were in mechanics. He formulated the laws of motion and a theory of gravitational attraction so universal that it could embrace both the movements of the planets in their orbits and the fall of the legendary apple from a tree. His mathematical and mechanical attitude naturally pre-

Alchemical symbolism

disposed him towards the prevailing 'corpuscular theory' of matter. He performed many experiments in chemistry and tried to explain them in a dramatic way using his new laws of motion allied to strong forces between the particles similar to the force of gravity.

> The particles of Acid . . . are endued with a great Attractive Force; in which Force their Activity consists; and thereby also they affect and stimulate the Organ of Taste and dissolve such Bodies as they can come at. . . . By their Attractive force also, by which they rush towards the Particles of Bodies, they move the Fluid, and excite Heat; they shake asunder some Particles, so much as to turn them into Air and generate Bubbles. (From *The Nature of Acids*.)

Attractive forces between the atoms were a new idea and very helpful in understanding how solids and liquids hold together and how new compounds are formed from simpler ones with the production of heat (motion).

By the time Newton wrote his great and most popular book, *Opticks*, in 1704, the Atomic Theory had become part of accepted thought. Gone were the days when to be an 'Atomist' was to be an atheist. Although Democritus, himself an avowed atheist, held that 'nothing comes about perchance but all through reason and necessity' he understood nothing of the mechanics of motion. His readers felt that the random movements of atoms and the collisions between them did reduce happenings to mere chance rather than to divine will. But Newton could point out the actual Laws of Motion. He was delighted to think that forces of attraction like those that rule the massive planets could also hold the tiny atoms together. He was fond of stating that 'Nature is very conformable to herself'; it strengthened his belief in God. He wrote 'Such wonderful uniformity in the Planetary system must be allowed the Effect of Choice' ie of a Creator. So the Atomic Theory became accepted, part of the body of thought of a believing age. It only remained for Newton to state

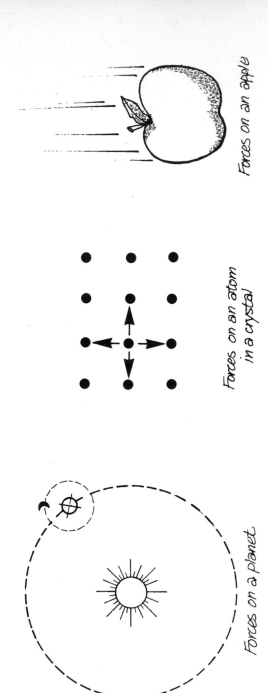

Forces on an apple

Forces on an atom
in a crystal

Forces on a planet

his belief couched in rolling phrases reminiscent of the great contemporary translation of the Bible:

> It seems probable to me that God in the Beginning form'd Matter in solid, massy, hard, impenetrable, movable Particles, of such Sizes and Figures, and with such other Properties, and in such Proportion to Space, as most conduced to the End for which he form'd them . . . so very hard as never to wear or break in pieces: no ordinary Power being able to divide what God himself made one in the first Creation. (*Opticks.*)

7
The Beginning of Chemistry

The seventeenth century had been most exciting and productive in science; experiments had become planned and much more accurate so that both old and new speculations as to the nature of matter could be methodically examined. Men like Boyle prided themselves on their scepticism towards the old orthodox theories, but they were vigorous and imaginative enough to put forward new ideas themselves. Newton agreed with Boyle and Bacon that the scientist's task was to collect facts and to abjure all 'occult theory'. He stated firmly in the *Principia* his famous dictum 'I frame no hypotheses'—but he belied it in almost every page of his second great work, the *Opticks*! Whether it was considered intellectually respectable or not, seventeenth-century scientists freely seasoned their experimental results with inspired hypotheses and the mixture was fruitful.

The following century was much more dogmatic in its approach. The feet of scientists and philosophers alike were firmly planted on the ground. The great French philosopher of the 'Enlightenment', Voltaire, appealed always to the 'facts of Nature' rather than to mere metaphysical thought. He had no interest in ideas which were not related to the realities of living—he called them *'romans de l'âme'*! Science was correspondingly popular and experiments were performed for the entertainment of high society in the fashionable salons of the day. This was the self-styled 'Age of Reason' when the English scientist and clergyman, Joseph Priestley, could preach sermons on the subject and include it in his religious books.

Above all things, be careful to improve and make use of the *reason* which God has given you, to be the guide of your lives, to check the extravagance of your passions, and to assist you in acquiring that knowledge, without which your rational powers will be of no advantage to you. . . . (*Institutes*, 1772.)

Yet, perhaps because of this very attitude, the period was barren of original scientific ideas.

Lavoisier and the Elements

However, collecting facts and specimens, the passion of eighteenth-century scientists, did help to get the new chemistry off to a good start. Antoine Lavoisier published his famous book *Traité élémentaire de Chimie* in 1789 as much to demonstrate his new system of naming the classifying substances as to report his discovery of oxygen or to give his tentative list of 'elements'. Although his work was immensely careful and, indeed, very important, the stultifying influence of the discipline of reason without imagination is clear to see. He declined to follow Newton's lead in considering the attraction or affinities between substances and he mentioned the word 'atom' but once in the whole book —and then in a context of disapproval! His guiding principle is made plain in the preface:

> . . . Imagination . . . which is ever wandering beyond the bounds of truth, joined to self-love and that self confidence we are so apt to indulge, prompt us to draw conclusions which are not immediately derived from facts.

How dry and discouraging his attitude seems today when compared with our wealth of imaginative speculation as to the origins of life and of the Universe itself!

Lavoisier clearly followed Robert Boyle's definition of an 'element' but was happier to refer to them more cautiously as 'simple substances'. Whether privately he pictured chemical reactions to himself on the atomic scale as the attraction and linking together of Newton's 'massy atoms' we cannot tell. Publicly, in his preface, he affirms his own rigidly empirical approach.

. . . if, by the term *elements*, we mean to express those simple and indivisible atoms of which matter is composed, it is extremely probable we know nothing at all about them; but, if we apply the term *elements* or *principles of bodies*, to express our idea of the last point which analysis is capable of reaching, we must admit, as elements, all the substances into which we are capable, by any means, to reduce bodies by decomposition.

The first section of his book is devoted to a discussion of heat (caloric) and to the discovery of oxygen and the new theory of burning. Only then does he give the following fascinating list of substances which, by the processes then available to him, he found impossible to decompose into yet simpler ones.

TABLE OF SIMPLE SUBSTANCES

Light	Antimony
Caloric (heat)	Arsenic
Oxygen	Bismuth
Azote (nitrogen)	Cobalt
Hydrogen	Copper
—	Gold
Sulphur	Iron
Phosphorus	Lead
Charcoal	Manganese
Muriatic radical (Chloride)	Mercury
Fluoric radical	Molybdena
Boracic radical	Nickel
—	Platina
Lime	Silver
Magnesia	Tin
Barytes	Tungsten
Argill (Clay)	Zinc
Silex (Silicates)	

This list is intriguing! It is the result of carefully recorded experiments by a man totally uninfluenced by the old alchemical ideas. To that extent his firm rejection of 'imagination' was of the greatest benefit. The only apparent debt to alchemy is the correct and impressive list of seventeen different metals that had been isolated. He was quite aware that his placing of 'earths' on the list was probably wrong but the appearance of 'light' and 'caloric' at the top of the table is more surprising.

Newton, as a mathematician, had always treated light as a stream of small projected particles. This enabled him to use his powerful laws of motion once again and to give mechanical explanations of reflection and refraction.

But Lavoisier was a chemist; he saw all change in terms of the combination of substances into more complex ones which, in turn, could be decomposed into the simple elements. It was principally because light could 'react' with the green substance in the leaves of plants to build up more compounds that he included it in his classification of the elements.

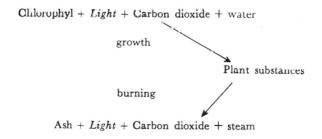

Chlorophyl + *Light* + Carbon dioxide + water

growth

Plant substances

burning

Ash + *Light* + Carbon dioxide + steam

Caloric (heat) was also treated as an element—a 'self-repelling' fluid which was combined with water in steam and locked into the chemical composition of fuels and explosives. We may now see both light and heat as forms of disembodied energy but such an idea was too sophisticated for the eighteenth century and indeed, in the twentieth century, Einstein's Theory of Relativity has blurred the clear distinction between energy and matter. So, for a time, Newton and Galileo's bold hypothesis that heat consisted only in the motion of atoms fell out of general favour and, to this day, we hold a

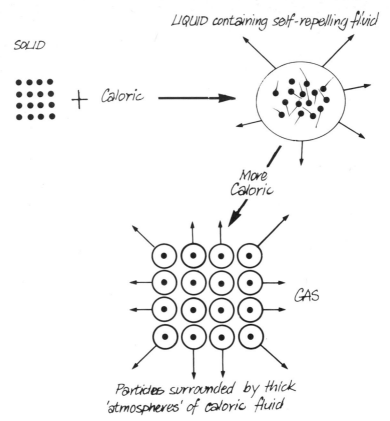

SOLID

LIQUID containing self-repelling fluid

+ Caloric ⟶

More Caloric

GAS

Particles surrounded by thick 'atmospheres' of caloric fluid

legacy of the ephemeral fluid theories of Lavoisier's times in our use of the word 'flow' for both heat and electricity.

The list also contains the three common gaseous elements—oxygen, nitrogen, and hydrogen. This was the culmination of a long century of patient experiment and slow understanding. Now gone, at last, were the vague 'airs' of earlier times and in their place were distinct and definite gases included as real matter on an equal footing with solids and liquids.

Simple experiments on the combining of gases had been performed by both Priestley and Cavendish with some interesting results. It took, for example, exactly twice as much hydrogen as oxygen for the synthesis of water. If the proportions of combining gases were in any other ratio and the mixture was 'sparked'—water was formed but there was extra gas left over. Other gas reactions also gave the same simple whole-number results, eg, 1 : 1, 1 : 2 or 1 : 3. It looks to us now as though this provided the very first confirmation of the truth of the Atomic Theory, but, logically, the argument could not yet be clinched. There was no reason to suppose that the different gases, each of which had different densities, actually contained the same number of particles, volume for volume. The age was far too 'enlightened' to jump into such an obvious logical pitfall!

8

The Atomic Theory comes of Age

By the beginning of the nineteenth century there existed in Paris a brilliant school of scientists and mathematicians who were mostly graduates of the newly founded *École Polytechnique*. Nevertheless the man to whom the honour of producing the first concrete evidence for the existence of atoms rightfully belongs was a hard-working Manchester schoolmaster—John Dalton. Born into a poor Quaker family his education at the local village school ended at the age of eleven, but his enthusiasm for science lasted all his life. His mathematics was never sophisticated but he was one of the most distinguished amateur scientists of all time. His early work was concerned with the gases and vapours of the atmosphere and he used the current atomic model to visualise the particles in a gas. Characteristically he could describe this theory with clarity, imagination, and even humour:

> . . . each particle occupies the centre of a comparatively large sphere, and supports its dignity by keeping all the rest, which by their gravity, or otherwise are disposed to encroach upon it, at a respectful distance. (*New System of Chemical Philosophy*.)

Like Priestley and Cavendish before him he had found that the same simple volume relations held when nitrogen combined with oxygen to form either the transparent gas nitric oxide, or the brown fumes of 'azotic' gas, nitrogen peroxide. In his notebook of 1803 he wrote:

> . . . oxygen joins to nitrous gas sometimes 1·7 : 1 and at other times 3·4 : 1.

Dalton collecting marsh gas (methane) by stirring up a pond to the amusement of his pupils, by Ford Madox Brown

The impurity of his nitrogen concealed the true values of the ratios which should have been 1 : 1 and 2 : 1, but the 'doubling-up' of the nitrogen in the two cases comes out clearly enough. It was probably this experiment which gave him the clue which he was shortly to follow.

The only convincing work could be on the combining weights of solid substances since gases were so little understood. However, even solid compounds had only just emerged by this time as having a definite and invariable composition. As late as 1797 the French chemist Proust had synthesised a compound known in nature and had shown, for the first time, that its composition was identical with the natural specimen. Even so there were many distinguished chemists of the time who could not accept that the same compound always had the same composition whatever its previous history. Mixtures, solutions, and alloys seemed so variable and many scientists were so wary of accepting any theory which, in its simplicity, seemed reminiscent of the now discredited alchemy, that they preferred to make no assumptions at all.

Dalton was clearly convinced of the truth of the atomic hypothesis and, as early as 1803, he had prepared some calculations as to the relative weights of the atoms of several different elements. It seems that he had been examining the *weights* of hydrogen and carbon in the two inflammable gases methane and ethylene. He found that the weight of hydrogen which combined with a fixed weight of carbon in the two cases was *exactly* doubled. To a scientist in the atomic tradition this was very promising! If it were not an isolated case, Dalton thought it could mean only one thing—that on the atomic level the compounds were formed by—

1 atom of carbon + 1 atom of hydrogen →   1 particle of ethylene

and

1 atom of carbon + 2 atoms of hydrogen →   1 particle of methane

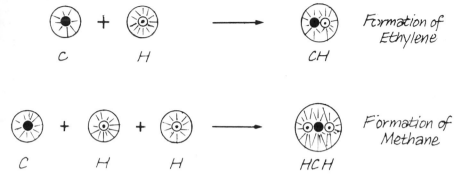

Further examination showed him that whenever two *different* compounds could be formed from the *same* two elements, the combining weights were always in the same simple whole-number ratios. Numbers like 1 : 1·789 did not ever occur, because only *whole* numbers of atoms could combine together to form compounds. Either one, two, or three atoms of one element with one or two atoms of the other.

Carbon monoxide and carbon dioxide, the two oxides of iron, and the acidic oxides of sulphur were obvious examples which could be investigated. Dalton's private ambition to measure the relative weights of the different atoms could now be realised. Instead of publishing a long list of experiments to establish the Atomic Theory he set out simple rules for himself and went ahead with his measurements. Nevertheless an important milestone had been reached and was generally recognised. Whether his figures for atomic weights were right or wrong (as inevitably many of them were.) the Atomic Theory had acquired *after two thousand years* its first piece of experimental evidence and was here to stay.

The Poetry of Science
By the beginning of the nineteenth century, science was climbing to a peak of popularity that it had never reached before, nor perhaps

since. This was the period of the Romantic Movement in literature and there was seen to be plenty of 'romance' in some branches of science too. The invention of the first electric battery by Volta was, indirectly, one of the influences that made Shelley's second wife, Mary, write the story of the monster of *Frankenstein*. Twenty years before that the most romantic of chemists, Humphry Davy, was lecturing and demonstrating on 'galvanism' to packed and enthusiastic audiences at the Royal Institution, in London. Davy himself wrote a great deal of romantic poetry and was a close friend of Coleridge and Wordsworth. They, in turn, came to see, in science, a way of glorifying and enjoying Nature complementary, in some ways, to their own. In his preface to the *Lyrical Ballads* (1802) Wordsworth wrote:

> Poetry is the breath and finer spirit of all knowledge; it is the impassioned expression which is in the countenance of all Science.
>
> and,
>
> If the time should ever come when what is now called Science, thus familiarised to men, shall be ready to put on, as it were, a form of flesh and blood, the poet will lend his divine spirit to aid the transfiguration, and will welcome the being thus produced, as a clear and genuine inmate of the household of man.

It was in this spirit that Coleridge attended so many of Davy's lectures on chemistry; in order, as he said himself, to renew his stock of images. Davy himself was invited by Wordsworth to punctuate and look over the proof sheets of his *Lyrical Ballads* before publication. But even if there had not been this close personal relationship between the brilliant young chemist and the great poet, what each was attempting in his own field would still have shown a striking similarity. Wordsworth was trying to present poetry in the simple language of the day at exactly the same time as Davy was lecturing and demonstrating in public his latest ideas and discoveries in chemistry. The age of the 'common man' was beginning.

Humphry Davy's earliest work with electricity followed immediately on the news reaching him of the discovery of the first electric battery—Volta's 'pile' in 1800. He was then only 22 years old and working in Bristol on the composition of 'laughing gas'. He turned, with characteristic enthusiasm, to this new branch of chemistry and investigated in turn its effect on water, the changes on its metal plates, and the nature of the liquid used in its construction. Six years later when he used electricity to isolate the two new metals, sodium and potassium, he already had clear ideas on how to improve on the original electric pile.

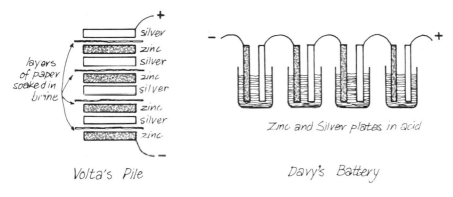

Volta's Pile

Davy's Battery

He felt instinctively that electricity could not have the power to generate matter, as some contemporary chemists supposed, and performed a painstaking series of experiments to prove that it merely decomposed pure water into its known elements, hydrogen and oxygen. Then he showed that the passage of an electric current through salt or acid solutions produced real movement of the metal or hydrogen part of the substance towards the negative side, and of the acid part towards the positive side. This was a brand-new kind of analysis—a separation of the parts of a substance by the simple application of electricity—and Davy predicted, at the time, that electricity would soon become a powerful tool with which to discover new elements. The substance of this work was delivered in Davy's first Bakerian lecture to the Royal Society in 1806. The

A contemporary cartoon by Gilray of a lecture by Humphry Davy at the Royal Institution. Laughing gas, one of Davy's earliest discoveries, is being administered to the fat gentleman

content and presentation made so striking an impact that it won immediate recognition and, despite the war between England and France, he was awarded from France a prize, inaugurated by Napoleon himself, for 'the best experiment on the galvanic fluid' in 1807.

The same year Davy used his electric battery to separate two new elements from the 'fixed alkalis' caustic potash and soda. These substances had been the object of many fruitless attempts at analysis, but Davy, now 28 years old, was confident that electricity was the key to success.

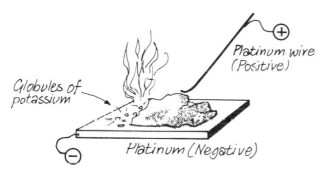

His cousin and assistant describes Davy's exuberant pleasure at the moment of victory:

> When he saw the minute globules of potassium burst through the crust of potash and take fire as they entered the atmosphere, he could not contain his joy—he actually danced about the room in ecstatic delight; some little time was required for him to compose himself to continue the experiment.

None of Davy's later work achieved quite the drama and excitement of this moment although he performed many more brilliant experiments and isolated several new elements in the next few years.

Does Electricity hold the atoms together?

Despite the fact that most of Davy's work was almost flamboyantly experimental he was also capable of very careful examination of a completely new idea in chemistry. What his work with batteries had shown him so clearly was that the forces which bind compounds together can be neutralised by electricity. It was natural, therefore, for him to suppose that these forces—the origin of the 'chemical affinity' between elements—were themselves electrical. For the last hundred years chemists had generally followed Newton in his opinion that these 'attractive forces' were like gravity. He had no experimental evidence for this view, he only held a belief, reflecting his religious attitude to the universe, that the same forces should rule all creation. Humphry Davy, on the other hand, set out to collect experimental proof to back his theory of electrical attraction. Electricity was still in its infancy, the battery itself was only a few years old, and there were neither ammeters nor voltmeters to measure current or potential. Armed only with a simple electroscope to detect

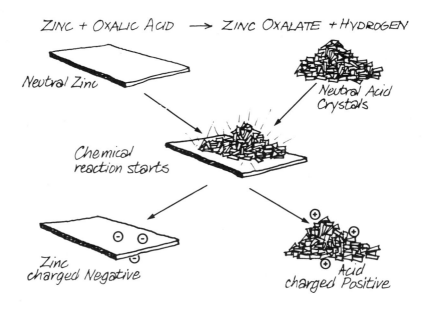

ZINC + OXALIC ACID ⟶ ZINC OXALATE + HYDROGEN

Neutral Zinc

Neutral Acid Crystals

Chemical reaction starts

Zinc charged Negative

Acid charged Positive

electric charges by the small movements of a fragment of gold-leaf, Davy examined chemical reactions in a completely new way.

Many metals, like zinc, react simply with acids. Humphry Davy used acid crystals, merely touched them with zinc and found at once that electric charges had been formed.

He followed this with a whole series of delicate and careful tests on alkaline reactions and simple heat reactions. He could show almost every time that opposite electric charges developed as the combination took place.

He knew already that many compounds, when in solution, contained oppositely charged parts which could be pulled apart by the electrical power of his new battery. There was enough material here for a hypothesis that as the elements combine together they acquire electrical charges and that this is what makes them hold fast together in the new compound just as large positively charged objects always attract those with a negative charge.

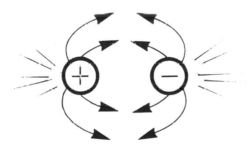

I drew the conclusion that the combinations and decompositions by electricity were referable to the laws of electrical attractions and repulsions; and advanced the hypothesis, 'that chemical and electrical attraction were produced by the same cause, acting in one case on particles, in the other on masses'. (1810—Ms for a lecture on Electrical Science, Davy.)

So Davy's direct and simple techniques launched a century of scientific progress that was to culminate in the discovery of an intimate connection between electricity and matter and of the existence of electric charges within the very atoms themselves.

Michael Faraday

In 1813 Sir Humphry Davy, as he then was, engaged a new laboratory assistant, a young man of 22 already bewitched by science and hero-worship of Davy. This was Michael Faraday who was to become one of the architects of modern electricity. He was a very different man from Davy, but although he had none of his master's flair for public speaking and demonstration he kept up the tradition of popular lectures at the Royal Institution and inaugurated both the Friday Evening Discourses and the Christmas lectures for children which have continued to this day. His was a more direct, less romantic, character than Davy's with an earnest, simple approach to both science and religion. He kept these two sides of his life in 'watertight compartments' as so many other Victorian scientists were to do. His faith was fervent and rigidly fundamentalist but, though he was always approachable and honourable to a degree, his experimental work was both disciplined and controlled. The most emotional entry in his scientific diary, after a brilliant series of experiments during which he invented the electrical generator, reads: 'Experiments with a single wire. Beautiful.'

In general he strove to remain rigorously logical in his scientific work,

> I must keep my researches really *experimental* and not let them deserve anywhere the character of *hypothetical imaginations*.

Faraday's early work, in conjunction with that of Ampère, Oersted, and Ohm, brought about a quantitative understanding of electric current, charge, and voltage. Using these concepts his experiments led him to invent the prototype of *both the electric motor and the dynamo*, so it seems extraordinary that the most common question he was asked by visiting industrialists of the 'steam-age' was 'What is the *use* of electricity?' Once while he was describing his theories, Mr Gladstone, then Chancellor of the Exchequer, interrupted him with the usual query and Faraday replied, 'Why sir, there is every probability that you will soon be able to tax it!' How depressingly contemporary a response!

In the 1830s electricity was still generally thought of in terms of the 'galvanic fluid'—Faraday was almost alone in his concept of fields of force surrounding both electric charges and magnetic poles.

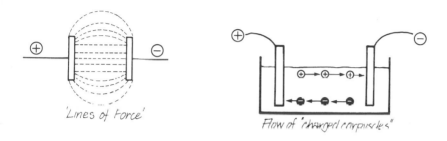

'Lines of Force'   Flow of 'charged corpuscles'

When, in 1833, he turned his attention to electrochemistry he was able to show that current flowed *only* when there was a chemical decomposition and that the strength of the current was *not* sufficient to cause the break up of dissolved substance. So it followed that this flow of current was due to 'internal corpuscular action'—the movement of charged particles (that he christened 'ions') which were already present in the solution. He visualised them moving along the lines of force that he had already mapped out. By careful measurements he could also prove that the *weight* of the ions that had arrived at the terminal had been neutralised and deposited on it, was unaffected by the length of its path through the solution, and depended only on the total *charge* which had travelled through the solution. Therefore it was these ions alone which carried the electricity.

Is Electricity atomic too?

The final series of experiments was designed to find out how the weights of the *different* substances deposited by the *same* current compared with each other.

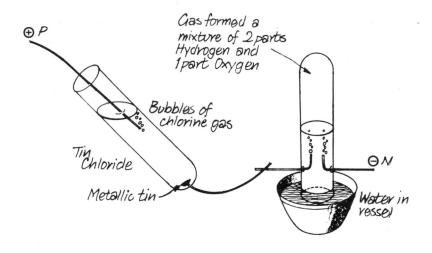

From sketches in his notebooks we can tell exactly what apparatus Faraday used for each of the dozens of experiments he performed. In his famous paper on Electrochemical Decomposition which he read to the Royal Society in 1834 he quoted the measurements he had obtained using *molten tin chloride*. When a definite volume of gas had collected in the second vessel he disconnected the apparatus and carefully detached the little knob of pure tin that had formed on the platinum wire in the first vessel and weighed it. He also calculated the weight of the gases that had been liberated. These were his results for the experiment:

Weight of tin formed = 3.2 grains

Weight of hydrogen and oxygen mixture = 0.49742 grains

These figures are in the ratio of 57.9 to 9 which is exactly the same as the normal combining weights of these elements. All his experiments gave the same conclusion—that the same electric charge was always carried by the weights of any two substances that would combine together. Clearly, when a compound was formed it resulted in the charges being exactly neutralised. Not only was this, as Davy had suggested, the basis of the 'affinity' which caused chemical reactions to take place, but *all* charged atoms, or 'ions', carry either exactly the same charge, double the charge or treble the charge. A generation before, John Dalton had shown (though not to everyone's satisfaction) that matter consisted of small discrete particles or atoms; now Faraday's work seemed to show that electricity too, when it was associated with matter, existed only in small discrete bundles.

This idea, like Dalton's, was very slow to win general acceptance. Indeed even Faraday himself seemed to have only half believed it:

. . . the results prove that the quantity of electricity which, being naturally associated with the particles of matter, gives them their combining power, is able, when thrown into a current, to separate these particles from their state of combination . . . the equivalent weights of bodies are simply those quantities of them which contain equal quantities of electricity. . . . Or, if we adapt the atomic theory or phraseology, then the atoms of bodies which are equivalent to each other in their ordinary chemical action, have equal quantities of electricity naturally associated with them. But I must confess I am jealous of the term *atom*; for though it is very easy to talk of atoms, it is very difficult to form a clear idea of their nature, especially when compound bodies are under consideration.

Michael Faraday was the last great scientist of the nineteenth century to withhold his recognition of the atom. His work ends the early doubting period of the new corpuscular theory.

An early photograph of Mr and Mrs Faraday

9
Speed and Strange Rays

During this time the Industrial Revolution had been changing the face of Britain, but it had little connection with the current work in pure science. The great steam machines of the new age were based on individual invention rather than on scientific research. There was no close relation between science and technology as there is today— no money from industry to found new laboratories nor to institute new research projects. The steam engine itself had been churning and clattering for more than a century before scientists could fit its output of power into the new theories of thermodynamics. It is true that the proliferation of railways made travel and exchange of ideas easier for scientists as well as for others. Perhaps the very notion of travel may have stimulated the many experiments to measure the speed of molecules, of light, and of other radiations in the latter part of the century. Certainly the idea of travel was taken up in the novels of Jules Verne—the forerunners of the Science Fiction novels of today.

Slowly the Romantic Movement thawed out the attitudes of the cold age of reason and made a new approach to science possible. Faraday was also the last scientist to abjure 'imagination' so severely; soon it was to be respected and used again as an essential part of the scientist's equipment as it had been in earlier times. Challenging fresh hypotheses were proposed more readily. Work in France on the wave-nature of light and the prediction of electromagnetic waves travelling in a weightless 'jelly' of aether certainly strained the imagination to a new extent! The fact that individual atoms and molecules (combinations of atoms) were not directly observable was no longer a stumbling block. The turmoil of ceaseless collisions between the molecules in a gas was more easily visualised and their speeds were calculated indirectly by the measurement of overall effects like pressure and density. By the 1870s physicists were rewarded by the discovery of a directly visible result of individual molecular movement. In 1828 a botanist, Robert Brown, had reported the continual agitation of microscopic particles suspended in a liquid. Fifty years later the effect was explained in terms of the bombardment of these minute objects by the impact of hundreds of the still smaller, fast-moving molecules.

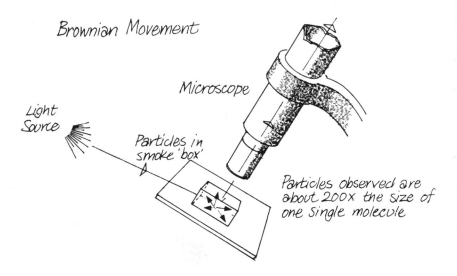

Brownian Movement

Microscope

Light Source

Particles in smoke 'box'

Particles observed are about 200x the size of one single molecule

This was a remarkable experiment which could also be observed when particles in smoke were viewed under a microscope. Brown referred to it, in bewilderment, as 'perpetual motion' and indeed the random jostling of these tiny bits and pieces by the unseen molecules is always bizarre and yet convincing. With the aid of observation, measurement, and imagination, molecules, the tiny 'building blocks' of matter, had at last won real acceptance. The next developments were to be much more weird.

The passage of electricity through air and other gases was more difficult to study. At ordinary pressures a vivid mauve spark could be seen, if the voltage was high enough, accompanied by a noise like a sharp crack. This is the familiar lightning and thunder of storms, but on a laboratory scale. When the pressure was reduced by a pump the electricity flowed quite quietly but the gas left in the tube began to glow with a characteristic colour exactly as we see in neon tubes today.

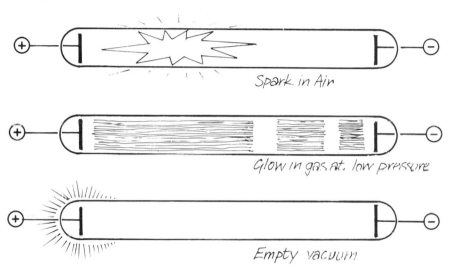

Spark in Air

Glow in gas at low pressure

Empty vacuum

So much was known in Faraday's time. As pumps grew better it became possible to obtain a flow of current in almost complete darkness where the only visible effect was a strange green fluorescence on the glass opposite the negative end. What sort of strange new radiation was coming from the cathode (as the negative plate was called) and causing this green glow, no one knew. This was the phenomenon that William Crookes set out to investigate in the 1870s.

It was already known that these 'cathode rays' could cast shadows so Crookes built a tube incorporating a star made of aluminium foil, which was placed in the path of the rays, and a very thin screen of uranium-glass which gave a bright yellow-green fluorescence. A hard black shadow of the star was thrown on the screen, always slightly larger than the actual object.

If the cathode rays had been similar to light rays the result would have been a blurred shadow, so another explanation of the nature of these rays was required. Crookes also noticed that, when he switched on the current his frail glass screen recoiled at the instant of impact as though these 'rays' were like a hail of tiny bullets. He decided to examine this property further.

In the windows of opticians' shops today it is quite common to see a glass bulb containing a tiny 'windmill' with four square metal sails, blackened on one side, which rotates slowly in the heat of a light bulb. This was also an invention of Crookes' which he called a 'radiometer'. He decided to use one of these, but without the black paint, to find out if the cathode rays really contained enough momentum to cause movement. He built another tube in which he suspended a radiometer so that it could be moved in and out of the shadow.

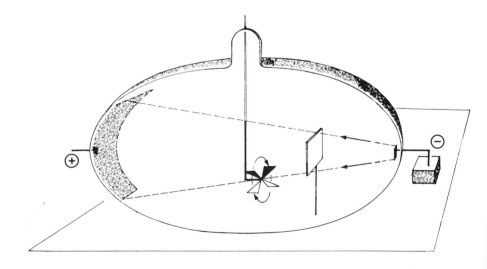

This experiment was a great success. When the cathode rays hit only one side of the little windmill it was rotated one way, when in shadow it stayed quite still, and when the rays hit it on the other side it revolved in the opposite direction.

Finally Crookes was able to show that these invisible rays—or particles as he now believed them to be—could be bent by holding a powerful magnet near the tube. He also demonstrated that the rays could be reflected by mirrors and focused on to a piece of metal which was heated white hot!

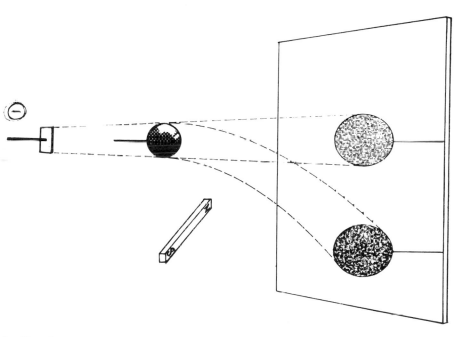

Crookes published his series of experiments in 1879. It only remained for him to present his own explanation of the nature of this odd new radiation in the customary fashion, at the end of his paper. William Crookes was not short on speculation being an ardent spiritualist and later a President of the Society for Psychical Research.

His conclusion shows an almost exaggerated awe at the strangeness of these rays.

> The phenomena in these exhausted tubes reveal to physical science a new world—a world where matter may exist in a fourth state, where the corpuscular theory of light may be true, and where light does not always move in straight lines, but where we can never enter, and with which we must be content to observe and experiment from the outside.

During the next twenty years more experiments were carried out with these mysterious rays and the controversy as to their nature continued. Crookes had come down in favour of their being 'charged molecules', but others held them to be waves like light. It was found that they could penetrate a sheet of aluminium, if it were thin enough, and that they seemed to carry a negative charge. This was evidence that they might be minute charged particles. In 1895 Röntgen discovered that they gave rise to yet another radiation—X-rays—but still no reliable measurements were available to clinch the argument. It was about this time that J. J. Thomson began his historic work on these cathode rays.

10

The Discovery of the Electron

From 1884 J. J. Thomson had been Professor of Physics at the University of Cambridge where he headed a brilliant group of young research physicists at the Cavendish laboratory. At this time university laboratories were a comparatively new development—the Cavendish had started in 1874—and from now on it was possible to have a whole series of experiments, following independent but related lines of investigation, going on 'under one roof'. 'J.J.' encouraged graduates from overseas to come and work at the Cavendish and the result was a very gifted group of research workers, almost all of whom were engaged in original experiments on some aspect of the conduction of electricity through gases. They stimulated each other, exchanged advice on apparatus, and encouraged beginners. No wonder, then, that their results had an accuracy and logical thoroughness which had been rare in science before.

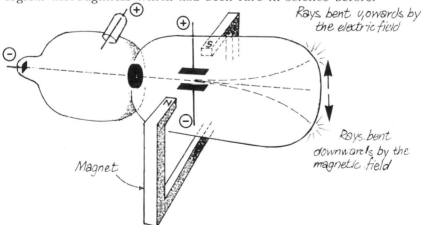

'J.J.'s' apparatus for 'weighing' cathode rays

'J. J.'s' own first experiments on cathode rays showed clearly that they did have a negative charge and that therefore the beam could also be deflected by an electric field. He set up a quantitative experiment, now regarded as a classic, to compare the bending of the rays by both magnetism and electricity.

The object of this experiment was to 'weigh' the particles or to find the size of the electric charge that they carried. Since the particles could not be induced to stand still but were only available as they raced at tremendous speeds across the near-vacuum inside the tube, the problem was not easy! By carefully measuring the amount that the beam was deflected and by 'balancing' the two fields so that the beam stayed quite straight it was possible to calculate only the speed of these corpuscles and the ratio

$$\left[\frac{\text{Mass of one particle}}{\text{The negative charge it carries}} \quad \frac{m}{e} \right]$$

Both these results were amazing! 'J.J.' found the speed of the particles to be thousands of miles per second and the ratio (m/e) to be less than $1/1,000$ of that of a charged hydrogen atom—the lightest atom known! He repeated his experiments with different gases in the tube and with cathodes made of different metals. It made no difference, the all-important quantity (m/e) remained exactly the same. Later he examined the radiation given off by red-hot wires and from sheets of zinc illuminated by ultraviolet light. Again and again he obtained the same value for (m/e)—the recognition-tag of cathode rays. 'J.J.' realised that a very fundamental new particle had been discovered.

Putting the results of all these experiments together, J. J. Thomson concluded that cathode rays consisted of negatively charged 'corpuscles' with the same (but opposite) charge as a hydrogen ion but more than a thousand times smaller (this would account for the way in which they shot straight through thin sheets of metal). He firmly suggested that they were a universal part of each and every atom

'J.J.' at work in the Cavendish

that exists. Not since the 'primordial element' of the early Ionic philosophers had there been any idea as simple and bold as this! But then 'J.J.' believed strongly in the value and power of ideas.

> Of all the services that can be rendered to science the introduction of new ideas is the very greatest. A new idea serves not only to make many people interested, but it starts a great number of new investigations.

He said this in 1909 of his brilliant pupil, Ernest Rutherford, but it was equally true of himself. He published his first work on cathode rays in 1897 and presented his astonishing new theory at the end of the paper in this simple paragraph :

> Thus we have in cathode rays matter in a new state, a state in which the subdivision of matter is carried very much further than in the ordinary gaseous state : a state in which all matter—that is, matter derived from different sources such as hydrogen, oxygen, etc —is of one and the same kind; this matter being the substance from which all the chemical elements are built up.

This idea, that the atoms themselves might be composed of much smaller pieces which were common to all the different atoms, was very exciting. It had exactly that grand simplicity, that evidence of the underlying unity of nature, for which scientists and philosophers had always been hoping. It is true that 'J.J.'s' theory met with some opposition but it started a blaze of experiments which were to convince everyone. The tiny charged corpuscles were christened 'electrons' and three different lines of research went on concurrently in the early years of this century.

Seeing the tracks of the electrons
One of J. J. Thomson's friends and colleagues—C. T. R. Wilson— had been staying at an observatory high up on Ben Nevis. The glorious cloud effects that he watched there influenced the direction that his research took for the next twenty years and gave physics

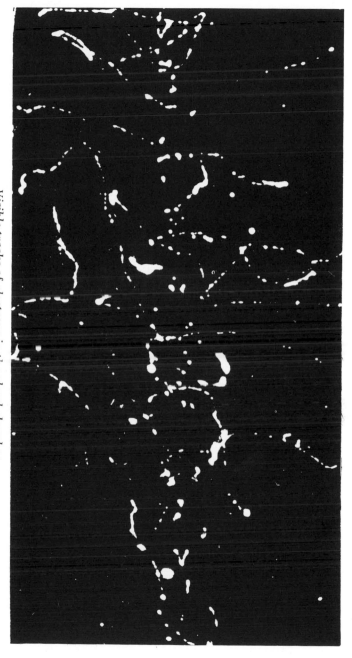

Visible tracks of electrons in the cloud chamber

a tool which was to prove endlessly exicting—a veritable hot-bed of new discovery! It was called a 'cloud chamber' and the principle behind it was to produce a cooled, super-saturated vapour which would condense along the track of any high-speed charged particle that entered the chamber, leaving visible vapour trails like those which high-altitude aeroplanes trace across the sky. In this way, for the first time, the path of a single electron could be studied.

Very shortly afterwards the same cloud chamber was used to observe the particles emitted from radioactive substances and the tracks of strange new particles out of 'empty space' which are now called cosmic rays.

Counting the electrons
J. J. Thomson, Wilson, and others tried to measure the actual size of the charge on the electron but it was not easy. Eventually an American, Robert Millikan, succeeded in making an accurate deter-

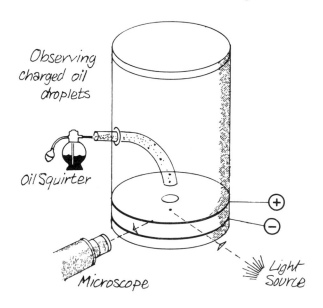

mination of the charge on tiny droplets of oil which he watched drifting slowly down across the field of his microscope. Those occasional droplets which captured a few free electrons from the air could be made to drift upwards by the attraction towards a positively charged plate in his apparatus.

Millikan examined hundreds of these droplets and obtained results similar to those in the following table:

(1)	14.4×10^{-10}	3 electrons
(2)	9.6×10^{-10}	2 electrons
(3)	19.2×10^{-10}	4 electrons
(4)	14.4×10^{-10}	3 electrons
(5)	9.6×10^{-10}	2 electrons
(6)	24.0×10^{-10}	5 electrons

Now these figures are not random. Each droplet's charge was either twice, three times, four times, or five times a basic unit of charge, viz, 4.8×10^{-10} electrostatic units (esu). Not only did this give an accurate value for the charge of a single electron, but it showed once more, and much more convincingly than Faraday had done, that electricity really did exist in discrete 'bundles'. One single unit of electric charge was the electron and no smaller bit is known to this day. The electron—the 'atom of electricity'—has never been split!

The positive part of the atom
Millikan also found positive charges on some droplets, of exactly the same size as the charge on the electron. Since electrons are to be found free in the air (especially when illuminated by X-rays), the atoms from which they have been stripped cannot be far away and they must each carry a positive charge. Naturally these two opposite charges must be such that they exactly balanced out when the original neutral atom was in existence. The problem was that, if the electron's weight was little more than 1/2,000 part of that of a hydrogen atom, where was all the atom's weight to be found? In

1906, when J. J. Thomson received the Nobel Prize for his discovery of the electron, he was already hard at work trying to examine the positive part of the atom.

Back in the same low-pressure discharge tubes that he had used before for the study of electrons, 'J.J.' knew that 'positive rays'

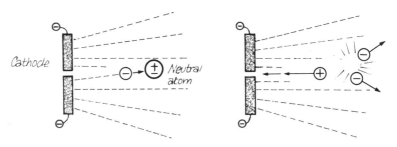

existed too. They always accelerated towards the cathode because of the attraction of its negative charge and if holes were bored in the cathode they would stream through it in dimly-glowing, coloured streaks. It became clear that it was the collision between a high-speed electron from the cathode with an atom or molecule of the remaining gas which knocked an electron out of the neutral atom and left it positive. These positive ions were much more difficult to deflect with a magnet than were electrons because they were so much heavier but 'J.J.' set out to measure the same ratio [mass/charge] for them too. It was a difficult experiment and he needed long exposure of a photographic plate to detect the various different ions in the tube but, in the next few years, he collected some interesting results:

1 The ratio (m/e) was different for each gas.
2 The mass (m) corresponded to the whole atomic weight of the gas.
3 Except for hydrogen it was possible to knock out one, two, three, or more electrons (up to eight for mercury vapour) from each atom leaving the ion with an equivalent high positive charge.

These results were the basis on which he built the very first model of the structure of the atom.

Thomson supposed that every atom consisted of a heavy, positively charged sphere in which sufficient light electrons were embedded 'like plums in a pudding' to make the whole electrically neutral. He was impressed by an experiment which showed how small floating magnets arranged themselves in circles around a powerful central electromagnet, and he imagined the electrons to be held in a similar way by the attraction of the largely positive atom.

This was only the first of several 'pictures' of the atom that the twentieth century was to produce. It was destined to be superseded within a very few years by a better model proposed by Thomson's former pupil, Ernest Rutherford, but 'J.J.' was always pleased to have had the 'first word'—who can ever know that they have had the 'last word' in science? Rutherford's new model of the atom grew out of the work on radioactivity which had been discovered by Henri Becquerel in 1896.

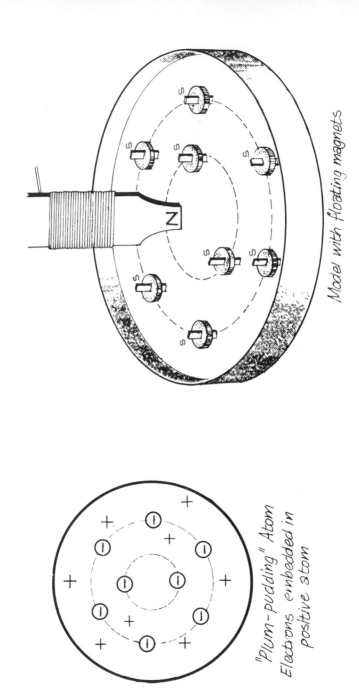

Model with floating magnets

"Plum-pudding" Atom
Electrons embedded in positive atom

II
Radioactivity and the Atomic Nucleus

While he was examining the phosphorescence of uranium salts, Henri Becquerel discovered that, even in the dark, they gave off a curiously penetrating radiation which 'fogged' a photographic plate although it was wrapped in both black paper and foil. At first this property seemed confined to uranium and its salts, but soon thorium was shown to behave in the same way. The famous work of Marie and Pierre Curie resulted in the isolation of two new radiating elements, polonium and radium. The latter was so 'radioactive', as Marie Curie called it, that it glowed visibly, produced a 'ghost' gas which was luminous in the dark, was continuously warm, burnt human skin, and finally destroyed itself. No wonder that many scientists were drawn to study it!

In 1899 Rutherford showed that two different radiations were produced by uranium which he called α and β rays. Becquerel proved that β rays were identical to electrons having the same (m/e) ratio. The α rays were much heavier and positively charged. While these particles were being shot out of radioactive substances curious changes seemed to be going on within the atoms themselves.

Rutherford teamed up with a chemist, Frederick Soddy, and together they examined the chemical nature of radioactivity. To their great surprise they found that the element thorium gave rise to a completely different new element, which they called Thorium X. It was even more radioactive than its parent. For the first time in history one element had been shown to be formed from another. How

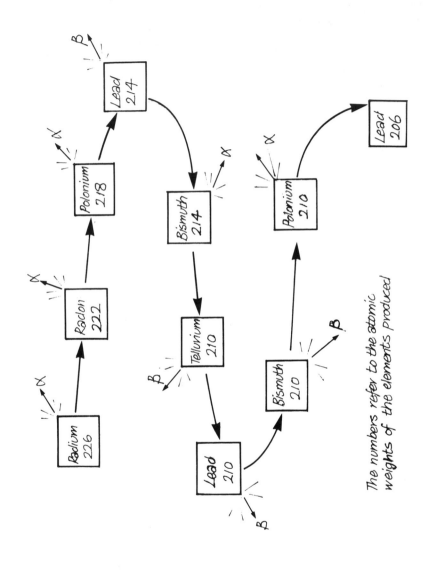

The numbers refer to the atomic weights of the elements produced

Lavoisier must have turned in his grave! They published their results in 1902, with a conclusion stating:

> Radioactivity is shown to be accompanied by chemical changes in which new types of matter are being continuously produced. . . . The conclusion is drawn that these chemical changes must be sub-atomic in character.

It was soon apparent that this conclusion was not confined to thorium but that every radioactive substance, as it emitted radiations into the surrounding air, was itself 'transmuted' slowly or quickly, into a completely different element. In the case of some radio-active substances whole chains of new atoms were produced, each one emitting energetic particles and changing into the next and so on until, at length, a stable atom was produced as the end-product.

The α-particle which was so commonly ejected by radioactive atoms had still not been definitely identified. Rutherford had shown that they were literally hurled out of the radium atom with speeds of about 15,000 miles per second, although they had a weight four times that of the hydrogen atom. Clearly the energy associated with radioactive atoms was enormous! With prophetic insight Rutherford wrote, in 1903:

> There is no reason to assume that this enormous store of energy is possessed by radio-elements alone. It seems probable that atomic energy in general is of a similar, high order of magnitude, although the absence of change prevents its existence being mani-fested.

Inevitably Rutherford received the Nobel Prize in 1908 for his work on radioactive decay. The following year he published his experi-mental proof of the nature of α-particles.

He used the radioactive gas, radon, that radium gives off, which was enclosed in a very thin glass tube. He knew that the α-particles could penetrate this and reach the vacuum in the outer vessel. He let them collect there for several days before he raised the level of the mercury to squeeze them into the top portion of the tube. Here an electric current was sent through the resulting gas and spectroscopic analysis of the coloured discharge showed, beyond doubt, that this was the light inert gas helium.

Therefore the α-particle emitted by radon, in common with so many other radioactive atoms, was simply a doubly charged atom of the element helium.

All atoms are made of the same pudding!
The facts of radioactive decay had shown the basic similarity of all atoms. If one atom could spontaneously break up, shooting out a fragment of itself which was helium and leave a residue which was quite another element, it must follow that all the different atoms were composed of the same matter in different quantities. It was as though the 'plum-pudding' atom of J. J. Thomson contained universal ingredients! A large lump of it was a heavy atom and a small lump a quite different lighter atom. The primordial elementary matter must be the same in the atoms of every substance.

Now the second model of the atom was about to take shape. In 1909 some experiments had been performed by Hans Geiger (who later invented the particle-counter that bears his name) and Ernest Marsden. They showed that high-speed α-particles from radon could be reflected off thin metal plates as well as penetrating through them. In the dark, with a simple microscope, they laboriously counted each scintillation on a fluorescent screen which signalled the arrival of an individual α-particle.

These minute bullets could have been expected to tear through the 'jelly-like' atoms of J. J. Thomson's mode with very little deviation. In practice they were found to bounce off in all directions. It was Rutherford alone who realised just how astonishing the actual scattering of α-particles was. He expressed his amazement and understanding of these extraordinary events in two of the most memorable sentences of modern physics:

It was quite the most incredible event that ever happened to me in my life. It was almost as incredible as if you had fired a 15-inch shell at a piece of tissue paper and it came back and hit you.

The nearly-empty atom

During 1911 Rutherford worked out his new 'nuclear' model of the atom. He suggested that the tiny atom [about 1/100,000,000 cen-

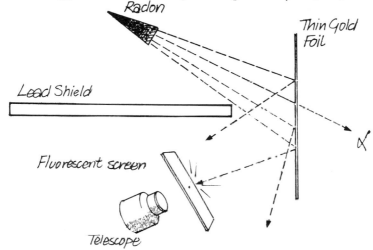

timetre across] was itself largely empty space, with all the positive charge and mass concentrated in a nucleus only 1/100,000 of the size of the atom itself. In this way he could calculate the probable number of α-particles which would be deflected at every angle by different atoms with their different amounts of charge. Further experiments showed very good agreement with all his predictions.

If all the positive charge was in the minute central nucleus of an atom, where were the negatively charged electrons to be found? Rutherford contented himself with a 'sphere of negative charge' round each atom, but in 1913, the Danish physicist Niels Bohr, who was at that time working in Manchester with Rutherford, completed the picture. He reasoned that the positive charge on the tiny heavy nucleus would attract the negatively charged electrons and swing them into orbits round the nucleus in the same way as the force of gravity pulls the planets into orbits round the sun. The different sizes of orbit would correspond to more or less energetic electrons, and if an atom was excited by heat or radiation the electrons might be 'lifted' to an outer orbit and then fall back, giving out a pulse of energy in the form of the coloured light characteristic of that particular element (yellow for sodium, green for copper, etc).

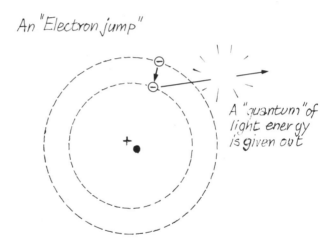

An "Electron jump"

A "quantum" of light energy is given out

Parking orbits for electrons?
The curious thing was that the definite lines in the spectrum of each element, and the new 'Quantum Theory' (see page 109), showed that only a very few of the infinite number of possible orbits were ever inhabited by electrons. There seemed no reason for this but Niels Bohr had to write it into his new model of the atom in order to fit the facts. He realised that his theory was an incongruous mixture of the old Newtonian mechanics and the new quantum theory (which postulated that energy, like matter, could only exist in definite discrete 'bundles'). However, its vindication came with his

precise calculation from theory of the wavelength of each line in the whole spectrum of incandescent hydrogen atoms.

> During the emission of the radiation the sytem [complete atom] may be regarded as passing from one state to another; in order to introduce a name for these states, we shall call them 'stationary' states, simply indicating thereby that they form some kind of waiting places between which occurs the emission of the energy corresponding to the various spectral lines. (From his paper 'On the Constitution of Atoms and Molecules', 1913)

Regarding the movement of a single electron the result was even more bizarre. If an electron could jump from one orbit to another and yet never be detected in any in-between position a new imaginative effort was required to picture it at all! (The only everyday analogy to this behaviour is the irritating way in which creases in a bed-sheet disappear in one place only to reappear in another as the sheet is stretched in a different direction.) This was the first indication that the 'elementary particles' of which matter is composed would behave in so odd a fashion that a whole new approach—quantum mechanics—would have to be learnt in order to understand them.

Not only did Bohr's model of the atom explain the visible spectrum of hydrogen but it was also successful when applied to the X-ray spectra of many different elements. When very fast cathode rays (electrons) impinge on a cold metal target X-rays are given out which are similar to light but of a wavelength a thousand times smaller. (Readers to whom these concepts are unfamiliar will find a simple explanation of them in Chapter 8 of the author's book *The Structure of Space*.) In 1914 another brilliant young pupil of Rutherford—Henry Moseley—measured the wavelength of these X-rays for a series of metals of increasing atomic weight.

On the quantum theory, radiation of such a short wavelength as X-rays must contain a great deal of energy. On the Bohr model this meant a long 'electron jump' into a deep orbit near the nucleus.

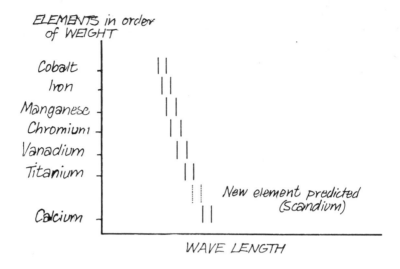

Moseley's results showed a very regular decrease in wavelength of the X-rays emitted as the weight of the elements increased. Mathematically it was not difficult to show that this constituted a proof of the steady increase in the positive charge on the nucleus, by one electronic unit for each successive element. Moseley's work was so fundamental that he could actually predict the existence of *three* unknown elements which were later confirmed and identified.

The 'grains' of the nucleus

Since the original work of Dalton the relative weights of the different atoms had been carefully measured. The early hope that they too might give a regular series where each atom weighed a whole number of times that of the hydrogen atom was apparently disappointed when atomic weights like those of the gases neon (20·2) and chlorine (35·5) were calculated. Unexpectedly this problem was cleared up by J. J. Thomson's work on positive rays. He discovered that when charged atoms of neon were deflected by electric and magnetic fields *two* distinct traces were obtained, one corresponding to an atomic weight of exactly 20 and another, fainter line, corresponding to an atomic weight of exactly 22. His pupil, F. W. Aston, refined the apparatus used and demonstrated that chlorine too was a mixture containing atoms of weight 35 and 37. He was able to prove that each variety or *isotope* of an element did, when taken separately, have an atomic weight which was an exact whole number multiple of that of hydrogen.

So two more simple 'whole number' rules had been established. The nucleus of every atom had a charge which was an *exact* multiple of the charge on a hydrogen nucleus and a weight which was also an *exact* multiple of the weight of a hydrogen nucleus. It was no wonder then that both Bohr and Rutherford leapt to the conclusion that the 'building blocks' of the nuclei of atoms were the hydrogen nuclei or 'protons' as they were called. Unfortunately a little juggling was required to make the numbers fit. The element helium, for example, had an atomic weight of 4 but a charge of only 2. To the four protons in its nucleus two negative electrons had to be added in order to neutralise two unwanted units of positive charge. The weight of an electron was known to be so small as to be almost negligible, so that this 'proton-electron' model of the nucleus fitted the known facts quite well.

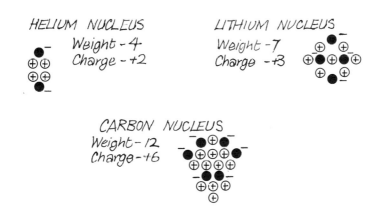

12
Atoms or Waves?

The outbreak of World War I was bound to disrupt the flow of scientific experiment and speculation. In Britain the most obvious results were the enlistment of most young scientists in the armed forces and the tragic death of many of them including the brilliant young Moseley. In Cambridge the Cavendish was almost deserted. Those who remained had to contend with OTC training every afternoon and 125 soldiers billeted in the laboratory. Hardly a favourable setting for research. The government made haphazard attempts to utilise the country's scientific talent for the 'war effort' and like most first steps it had its ridiculous side. J. J. Thomson served on the 'Board of Invention and Research' whose thankless task it was to examine and test the flood of wild suggestions sent in by members of the population who felt inspired to invent war-winning gadgets! At one stage these reached an average of 500 a day and included such gems as hanging a rope smeared with bird-lime from a balloon to entice and capture German zeppelins! Rutherford, W. L. Bragg, and others were more usefully employed on war-work connected with the detection of submarines by picking up underwater sounds. When the war was over scientific establishments under the aegis of the government or industry began to flourish. Though the prime object of these centres was applied or directed science, there have been many cases where so much latitude was allowed in research that as valuable contributions to scientific theory have come from these more prosaic establishments as from the 'ivory towers' of university science.

During the war there was also very little exchange of scientific literature between Britain and Germany, and in the immediate postwar years it became clear that a new spirit was abroad in physics which had its roots on the continent rather than in England. For twenty exciting years British physicists had led the world in the practical exploration of the atom. The experiments they devised were brilliantly simple and fundamental so that the rate of progress was staggering. Their mathematics and imaginative powers were correspondingly simple and direct—exactly what the early experimental period in a new branch of science required—but by 1913, even in its inception, the Bohr theory of the atom was antiquated. It was intellectually pleasing, even poetic, to see within the tiny atom a reflection of the gigantic solar system with electrons revolving around the nucleus like planets around the sun. It was just the sort of theory to please the British scientists, still immersed as they were in their long Newtonian dream of simple forces and defined orbits. In fact it pleased so much that now, more than forty years after its inevitable downfall, we still use phrases like 'planetary electrons', 'orbits', and 'electron spin'. Even without the tragic hiatus of World War I the 'new physics' was bound to have broken down the simple pictorial approach of Rutherford and Bohr because their theory was neither truly self-consistent nor did it explain all the known facts. On the classical theory an electron could not circle a positive charge without continuously radiating energy and spiralling in towards the centre. There was no conceivable reason why certain orbits should be possible and others impossible, nor why an electron should leap, in so sudden a fashion, from one to another. Finally neither Bohr nor anyone else could make his theory fit the spectra of any of the ninety-two elements, other than hydrogen. New inspiration was required and during the 1920s it slowly emerged.

Quanta—the ultimate pulses of light
During the war years, which were so barren in England, two 'giants' of theoretical physics were hard at work in Berlin. These were Max Planck and Albert Einstein. At the beginning of the century when physics was riding high and confident on the established wave-theory

of light, Planck had shown that light energy, in common with heat energy, seemed to be delivered in small discontinuous 'pulses' which were called quanta of energy or photons. He had evolved this novel, revolutionary theory not by an imaginative leap in the dark to explain the emission of radiation from hot bodies but by exploring every mathematical approach in turn. When at last he used a statistical method, the form of solution not only fitted the obdurate facts of experiment but also introduced this discrete entity—the tiny, indivisible quantum of energy. Low-frequency radiation like heat consisted of small quanta but the higher frequencies—visible light, ultraviolet rays, and X-rays consisted of quanta with larger quantities of energy. It is not hard to see why this theory was so indigestible for physicists of this time. The previous century had shown that light could 'spread out' like a wave when it passed through a narrow slit—how could 'pseudo particles' like light quanta do this? More convincing still, light waves could be made to interfere with each other. This means that though two waves which are 'in step' with each other add up normally, to give a big wave, if they are 'out of step' they will exactly cancel out. This effect can sometimes be heard with sound waves in a badly designed concert hall. Reflected sounds react with the original sounds to make some notes too loud and to subdue others. There seemed no way that photons could fit this phenomenon. They were to be visualised as 'pieces' of energy so that the time-honoured arithmetic must hold where $1 + 1 = 2$ but $1 + 1$ is never $= 0$! No wonder the quantum theory was so uninviting.

International Confusion

In 1911 a celebrated international conference on physics was held, with all the leading scientists present, principally to discuss the problem of the quantum. Max Planck, who had originated the theory, was there and so too was Albert Einstein. Among all the brilliant theoretical physicists and mathematicians who attended the conference only Einstein would maintain that quanta were *real*. The twentieth century had produced a new dilemma which was to haunt

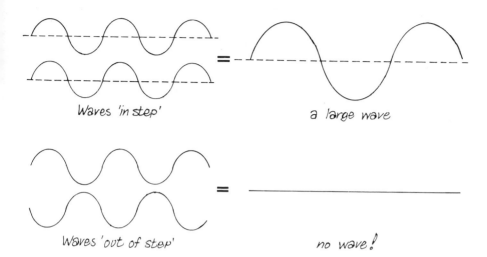

Waves 'in step'   a large wave

Waves 'out of step'   no wave!

it for several decades—there was mathematical expertise and experimental 'know-how' of a high order, but when physicists left their equations and their benches to talk together they were faced with the task of interpretation and understanding which was harder still. Mathematical abstractions have to be understood at a very basic level if they are to become physical realities. Max Planck himself found the 'quantum question' deeply worrying.

> . . . in physics . . . one cannot be happy without belief, at least belief in some sort of reality outside us. This undoubting belief points the way to the progressing creative power. A research worker who is not guided in his work by any hypothesis, however prudently and provisionally formed, renounces from the beginning a deep understanding of his own results. (Max Planck.)

Most of the scientists of this time clung to the more familiar Wave Theory of light and used the quantum theory uneasily 'with their eyes shut' when the problem compelled them to.

Two vibrating prongs produce sets of ripples which interfere with each other, leaving patches of calm undisturbed water where there are no waves at all

> Sir William Bragg was not overstating the case when he said that
> we use the classical theory on Mondays, Wednesdays, and Fridays,
> and the quantum theory on Tuesdays, Thursdays, and Saturdays.
> Perhaps that ought to make us feel a little sympathetic towards the
> man whose philosophy of the universe takes one form on weekdays
> and another form on Sundays.

This quotation from Sir Arthur Eddington's Gifford lecture delivered
in 1927 has a familiar ring to it. Here in twentieth-century physics
was a specifically twentieth-century problem. The conference in 1911
did little to elucidate the mystery. Einstein advocated imagining the
quanta being carried by a 'ghost wave of probability'—a prophetic
suggestion, but one that seemed almost mystical at the time.

One physicist at this famous conference was Maurice de Broglie
who later described the problems that had been raised to his younger
brother Louis, who was then only 19. The young man was so in-
trigued and excited by the conflicting theories that it was to become
the conscious background for his later amazing theory of matter.

On the other hand—could matter be waves?
By 1920 when Prince Louis de Broglie was a professor at the University of Paris, the Bohr theory of the atom was becoming more and
more suspect. As a theoretician de Broglie chiefly objected to the
'stationary states' or 'quantised orbits' of the electron which seemed
entirely arbitrary and without real explanation. Fascinated as he was
by the dual 'Wave-Quantum' nature of light his mind was open to
a parallel revolutionary theory of matter.

It is well known that stationary waves or vibrations of any kind in
the world around us can only fit into certain lengths. This is true
of vibrating strings or of resonating air in a bottle disturbed by a
tuning fork. By an inspired analogy de Broglie saw that if the *electron*
were considered as a *wave* the 'stationary states' of the Bohr atoms
would arise naturally as the wave fitted itself around he nucleus of
the atom.

The first paradox of the theory was the extraordinary result that slower particles were linked to higher velocity waves and that, in any case, these travelled faster than light. However, credibility was restored when de Broglie showed that it was not the continuous smooth wave that signified the presence of matter but the disturbance in it. Just as the progress of the meal down the digestive tract of a cobra may be determined by observing the movement of the 'lump' down its body; so the extent of the 'rough patch' in the waves of matter show the location of the more familiar particle.

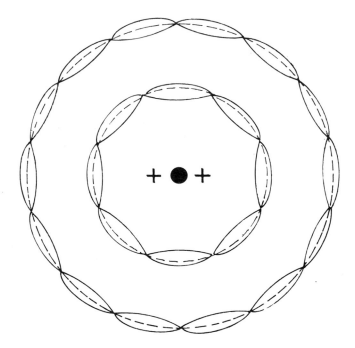

Two possible 'orbits' into which a whole number of stationary electron waves can fit

The best example of this is the wake of a boat. Here the 'disturbance' can be seen to travel quite slowly through the water with ripples

passing through it and dying out in front at a much faster speed.
The reader can look for this effect at leisure in the bath by watching
the circular disturbance that spreads through the water from an
occasional drip of the tap!

It is not surprising that, when de Broglie first delivered a paper
on this subject in Paris he was greeted with bewildered astonishment.
In spite of the sense his theory made of 'quantised electron orbits'
in the atom it might have become no more than a scientific curiosity
had no actual experimental evidence for his strange matter-waves
come to light in the next few years. De Broglie's pioneer work
sparked off the most crowded and creative five years in the history
of theoretical physics. Old ideas were stood on their heads and bizarre
new feats of imagination were required to understand the new ones.
Only recently, perhaps some thirty years later, is the turmoil settling
down as a new generation of physicists, who have grown up with
'matter waves', take them in their stride. However, by 1929 it was
already clear that de Broglie's synthesis was here to stay and he
was awarded the Nobel Prize. He concluded his address on that
occasion with these words :

> We thus find that in order to describe the properties of Matter
> as well as those of Light, we must employ waves and corpuscles
> simultaneously. We can no longer imagine the electron as being
> just a minute corpuscle of electricity : we must associate a wave
> with it. . . . It is therefore on this idea of dualism in Nature between
> waves and corpuscles, expressed in a more or less abstract form,
> that the entire recent development of theoretical Physics has been
> built up, and that its immediate future development appears likely
> to be erected.

13
The Struggle to Understand

In fact the old Bohr atomic model had had to become so complex in order to account for all the fine lines in the spectra of the elements that it had given rise to dissatisfaction among other physicists too. When elliptical orbits round the atom were called for other 'quantum numbers' had been postulated to describe the limited number of permitted shapes of these ellipses. Experiments with free atoms showed that the individual 'spinning' electrons could only orientate themselves in distinct directions in space. This so-called 'space quantisation' was an observable fact but hard to integrate into a classical idea of particles. When an old theoretical model gets so cluttered up with additional 'rules' designed to explain experimental results but not based on real understanding it is time to look for another theory!

In the years 1920–26 Erwin Schrödinger, then professor at the University of Zürich, was struggling with this very problem. His early work had been on the patterns of vibrations that can exist in three dimensions and he hoped to apply this approach to atomic theory. In two dimensions it was well known that many different patterns of vibrations could arise when a flat sheet of metal was vibrated. Indeed it used to be possible to hear street musicians in central London earning their living by playing tunes with a violin bow on a common saw-blade.

When Schrödinger heard, through Einstein, of de Broglie's theory of Waves of Matter it gave him his final clue. With more sophisticated mathematics he worked out the wave-equation for an electron in the field of force of an atomic nucleus, and demonstrated triumphantly that almost all the 'quantum numbers' arose naturally from it. No longer was it necessary to imagine the electron leaping instantaneously through forbidden space from one permitted orbit to another. Now, with the aid of the New Wave Mechanics, it was possible to calculate the shapes of the standing waves of an electron about the atom and to understand how, when the atom was 'excited' to a new energy level, one pattern would change into another.

Some possible shapes of the electron cloud in and about an atom, according to E. Schrödinger's theory

Experimental evidence of matter waves
Almost simultaneously with the appearance of Schrödinger's Wave Mechanics, experimental results turned up by accident which lent

great weight to the wave theory of matter. Two physicists, C. J. Davisson and L. W. Germer, working in America, were shooting electrons at a nickel target in a vacuum tube when the tube burst. They took out the piece of nickel, heated it to drive off any absorbed air, and allowed it to cool down very slowly before placing it in another tube. Without realising it they had provided exactly the right conditions for the metal to crystallise. When they continued their experiment in the new tube they obtained some baffling results. At some angles they could find almost no reflected electrons, though at other angles there was a strong reflected beam.

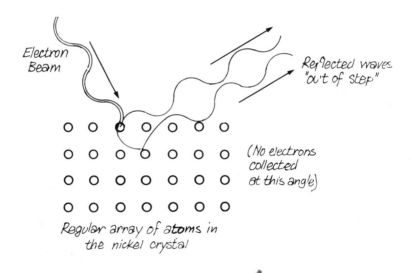

After reading de Broglie's work they realised that this could mean that, at certain angles, the electron-wave reflected from the top layer of atoms was 'out of phase' with that reflected from a deeper layer of atoms, and that *interference* was taking place. This result was well known with X-rays and had always been taken as proof of their wave-nature; but no one had ever thought of looking for such an effect with 'particles' like electrons. Careful measurement of the angles involved gave a figure for the wavelength of the electrons used and it was found to be exactly what de Broglie and Schrödinger had predicted. Nothing in the classical mechanics of particles could possibly explain this result. Now the conclusion was forced on the world of science that Wave Mechanics could no longer be considered as just an alternative, and more curious, way of looking at the world. It had a reality and power actually greater than that of the more comfortable, conventional view of the particles of matter. One plus one no longer equalled two!

More experimental work followed on this, including some with very fast electrons performed by J. J. Thomson's son—G. P. Thomson. Then heavier particles like hydrogen and helium nuclei were used. In every case it was found that the wavelength of the particle was largest when the mass and speed were low, and shorter when these were great—just as Wave Mechanics had predicted. This result has had practical applications, chief among which is the electron microscope which is widely used in biology and medical research for examining objects too small to be seen with ordinary light waves.

What does it mean?
The basic problems of the new Wave Mechanics did not lie in the laboratory but in the imagination and understanding of its adherents. Schrödinger himself was very unsure about the reality of the waves of matter. He thought that the amplitude, or size, of the wave was to be taken as a measure of the density of electrons at that point. However, the mathematical expression which he had obtained was 'imaginary'—that is it contained the square root of a negative number—which is clearly 'impossible' and 'unreal'. Secondly the medium in which the ripples of matter-waves moved and had their being (which he called 'q-space') seemed to be able to hold only *one* electron! In what sense could each electron exist in a separate 'space-time' and yet interfere with each other? The following passage, taken from a lecture of Schrödinger's on Wave Mechanics in 1928, shows his predicament at this time:

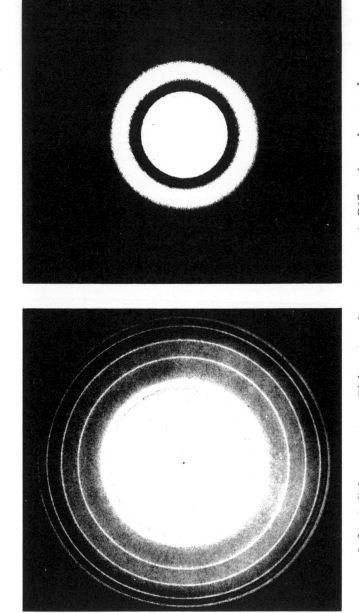

Left: Is light quanta? or, Right: Are electrons waves? Diffraction photographs comparing the behaviour of light going through a pin-hole, and electrons going through the holes between atoms

The statement that what *really* happens is correctly described by describing a wave-motion does not necessarily mean exactly the same thing as what really *exists* is the wave-motion . . . (Q-space). . . . Though the latter has a quite definite physical meaning, it cannot very well be said to 'exist'; hence a wave-motion in this space cannot be said to 'exist' in the ordinary sense of the word either. It is merely an adequate mathematical description of what happens.

It seems from this curious passage that Schrödinger is trying—in carefully chosen words—to claim that his mathematical fiction, though unreal, actually works! This careful, almost fussy, use of words was a by-product of the new synthesis of science and philosophy. Gone were the days of metaphysics and the pursuit of ideal concepts of 'truth', 'beauty', and 'goodness'. Perhaps the advent of psychology had something to do with this. Philosophers had been very impressed by the achievements and methods of science and they set out to analyse the use of language and its meaning more rigorously than ever before. Bertrand Russell examined the logical basis of mathematical concepts, and the 'Vienna Circle' of logical positivists, who flourished in the first twenty years of this century, tried to do the same for science. The great philosopher, Wittgenstein, wrote simply 'whatever can be said at all can be said clearly' and it seemed an excellent criterion to the logical positivists. Not only must language be unambiguous and 'meaningful' but, as in science, all propositions should be *verifiable* and the concepts used should be *observable*. Twentieth-century philosophy has continued to use these critical tests and made some interesting discoveries in logic and mathematics. Science, too, became more self-conscious and the next formulation of quantum mechanics showed a debt paid back from philosophy to science.

Werner Heisenberg was only twenty-four when he formulated an expression of quantum mechanics which is still widely used to this day. He had been a brilliant mathematician in school but had also read Greek and modern philosophy avidly in his spare time and this proved to be a vital aid to his scientific work throughout his life. In 1925 he was doing work on the scattering of light by gases— exactly the sort of problem that would involve the modern paradox in which light 'waves' will behave like a stream of 'quanta' and the 'solid' atom will be excited to different 'levels of energy'. Instinctively Heisenberg took the most logical approach which cut right to the heart of the problem. He decided simply to ignore the elusive electron and to use an array of numbers to describe the atom which contained only those frequencies of light that it could radiate since these, not the mythical 'orbits' of the electrons, were strictly *observable*.

In classical terms this seemed to mean that the electron might be in all possible 'orbits' at the same time, but to Heisenberg it meant simply that, since the electron could not be observed within the atom, it was both nowhere in particular and everywhere at once! Heisenberg always held that the phrasing of the question was as profound as the answer to be given. This is not really a difficult point to understand since science is concerned with real measurable quantities so its language should be exclusively about these. If you ask a silly (or unscientific) question, you will get a silly (or unscientific) answer!

The mathematics of these arrays of numbers was not easy. In his very first paper Heisenberg was not yet clear what operation to perform but he had already noticed a numerical peculiarity of his theory that was very new and basic. If p represents the momentum of a particle and q its position then—

$$[p \times q] \text{ is not equal to } [q \times p]!$$

It was his professor at Göttingen, the eminent Max Born, who read the paper and, after racking his brains for a troubled eight days, recalled a mathematical method that he had learned years before for dealing with arrays of numbers—Matrix Theory—which gave just this result. The following year Born, Heisenberg, and Jordan collaborated together in the mathematical formulation of this new quantum mechanics. It did not seek to answer the question 'Is matter *really* waves?'—it did not have to postulate a vibrating, rippling 'q-space', yet Schrödinger was delighted to be able to show that it gave exactly the same results as his own Wave Mechanics.

The Uncertainty Principle

The apparently ridiculous result that $(p \times q)$ is not the same as $(q \times p)$ meant, in terms of experimental observation, that to measure the momentum of a small particle, or quantum, and *then* its position would give a different result from first observing its position and *then* its momentum. Particles like electrons are so small that there is no way of observing them that does not either alter their speed or knock them out of position. Observing them at all must involve illuminating them, and the impact of a single photon of light will constitute very 'rough handling' to such a small particle. Heisenberg's concentration on scientific measurement had exposed a basic uncertainty, an ineradicable error due to the coarseness of nature itself, in all experiments which set out to measure simultaneously two quantities of one tiny particle. This is called 'the Uncertainty Principle'—a fascinating idea with a fascinating title!

During the winter and spring of 1927 physicists all over Europe wrestled with the difficulties of this new approach. Not only was there an ingrained uncertainty in all measurements but the nature of matter on this tiny scale seemed finally to defy imagination. Heisenberg was now in Copenhagen with Niels Bohr and he wrote of this period:

> I remember discussions with Bohr which went on through many hours till very late at night and ended almost in despair; and when at the end of the discussions I went alone for a walk in the neighbouring park, I repeated to myself again and again the question: Can nature possibly be as absurd as it seemed to us in these atomic experiments? (*Physics and Philosophy.*)

Max Born had shown that the real connecting link between Schrödinger's Wave Mechanics and Heisenberg's Quantum Mechanics was that they both calculated the 'probability' of a particle being found at any place. Gone was the comfortable certainty of previous ages; it seemed now that the most probable path of an electron was no more than a matter of statistics, and that the power of exact prediction of future events from scientific laws was lost for ever. Einstein, an older and perhaps a more religious man, exclaimed in horror, 'God does not play dice!' but the younger scientists struggled on until they could present the International Conference in the autumn of 1927 with a workable interpretation of the 'New Physics'.

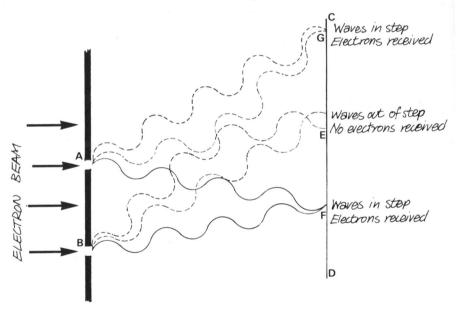

Consider a difficult experimental result that the 'Copenhagen School' had to explain. If a beam of electrons is allowed to fall on a plate with two very fine slits cut in it (A and B) they will spread out like ripples and, by the time they reach the screen CD they will form an interference pattern. There will be places on the screen, like E, where the two waves will be out of step and no fluorescence will be caused, and other places like F and G where the waves will reinforce to make a bright patch on the screen. The resulting dark and light pattern is the calculable probability distribution of Wave Mechanics and exactly similar to that obtained with all kinds of waves. However, if a fluorescent material is used which is sensitive enough to record the arrival of a single electron by a tiny scintillation of light then the particle nature of electrons becomes more apparent. Nevertheless when the electrons are allowed through the apparatus so slowly that they are recorded on the screen one by one and the total is counted up—still the same pattern is obtained. Now if an electron is a single particle it can only come through one slit; how then can it interfere with itself and never arrive at E but only at such places as F and G? Only *waves* can be 'out of step' with each other but they can go through two holes at the same time to cover the whole screen—only *particles* can produce individual scintillations but each can only come through one hole! Neither view is wrong, each is complementary to the other, but at this point the Uncertainty Principle must be invoked. The sudden scintillation on the screen gives precise information about the position of the electron at one moment in time. We must expect to pay for this accuracy by un certainty elsewhere and the separation of the slits, A and B, is exactly the measure of the predicted uncertainty using Heisenberg's Principle. It is quite impossible to 'pin down' the track of the electron more exactly and, in the spread of its 'probable' path, it can be said to have been as likely to have come through either of the slits or, indeed, both! Hence the interference pattern.

14
The Other World of Anti-matter

A young English mathematician of genius, Paul Dirac, produced in 1928 the final formulation—up to now—of the Quantum Mechanics of particles. He used Schrödinger's wave equation, but 'modernised' it to include Einstein's Theory of Relativity. He used Heisenberg's Uncertainty Principle, but applied it to *all* measurements of the electron, realising that every precise observation made on it influenced the electron to jump into some other state. The equations that he derived were mathematically elegant but, in one way, even more bewildering, Dirac had had to include the square of the energy of the electron and, as every school child knows, square roots can be both positive and negative. Electrons with positive energy are 'excited' or fast moving; electrons with little or no energy may be at rest—but what could 'negative energy' possibly mean? Dirac's first hope was that this inconvenient result could, ostrich-like, be ignored but the dictates of his superb mathematics would not allow it. By 1930 he had some tentative ideas on the problem.

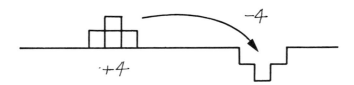

If negative numbers can ever mean something to us it is because they are the *opposite* to positive ones. To every number, however large, there is always an equal but opposite negative number that could exactly swallow it up. When Dirac investigated the possibility of electrons with negative energy which had cropped up so awkwardly in his theory he realised that they, too, were like 'holes' which could swallow up and actually annihilate an electron with the same positive energy! Of course nothing like this had ever been observed in nature but the mathematics of his new quantum mechanics was inexorable.

> These holes will be things of positive energy and will therefore be in this respect like ordinary particles. Further, the motion of one of these holes . . . will thus correspond to its possessing a charge + *e*. We are therefore led to the assumption that *the holes in the distribution of negative-energy electrons are the protons*. When an electron of positive energy drops into a hole and fills it up, we had an electron and proton disappearing together with the emission of radiation. ('A Theory of Electrons & Protons', *Proc. Roy. Soc.* 1930. Dirac.)

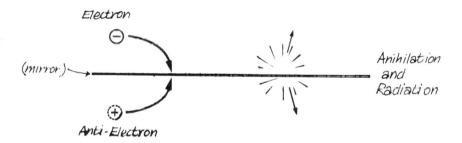

The only particle which was known at that time to have the same charge as the electron, but positive, was the 'proton' or hydrogen nucleus—but it did not really fit the part of an 'electron hole' as it was so much more massive than the electron. Within two years a new particle—the 'anti-electron'—was to become an experimental fact. This is one of the most astonishing examples of brilliant, though reluctant, prediction from theory.

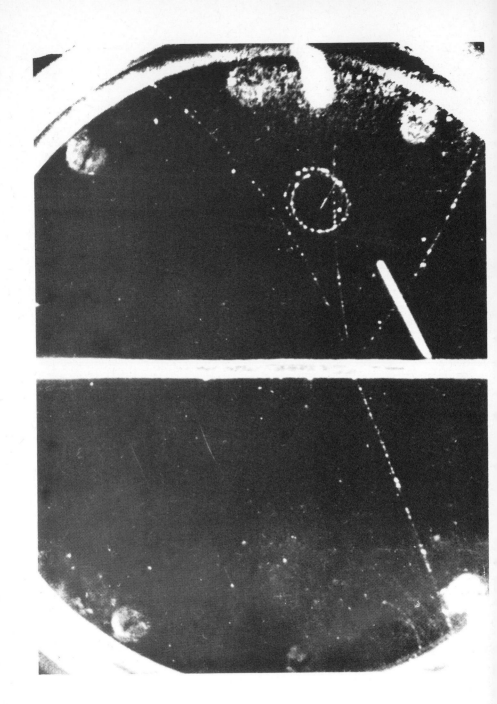

Finding the 'holes'

Ordinary, negatively charged, electrons are very common in our world but electrons in 'negative states' are bound to be vanishingly rare since they behave like holes and will rapidly be 'filled up' by ordinary electrons. An enormous amount of energy will then be required to lift the electron out again and so make both it, and the empty 'hole', detectable. This sort of energy was known to be available in the γ-radiation from outer space, called cosmic rays. These had been studied since 1912 when observations high up in balloons had been made, but their energy is such that their effects are easily observed at ground level—or, indeed, deep underground!

The best apparatus for recording the tracks of individual particles was still the Wilson Cloud Chamber and it was such a 'vapour trail' that first demonstrated the existence of the anti-electron. In 1932 C. D. Anderson, a pupil of the same Robert Millikan who had originally measured the charge on the ordinary electron, photographed the first known tracks of an electron with the 'wrong' charge on it. Its path was forced into a curve by a magnetic field and the piece of lead was present to slow down the particle and verify that it really was travelling from bottom to top and was not an ordinary negatively-charged electron moving in the opposite direction. Scientists in the Cavendish rapidly took up the search and soon demonstrated that each time a high energy cosmic ray produced a positive electron, a negative electron (curving the opposite way) was also formed. However, whereas the ordinary electron spirals gradually to rest, the strange anti-electron often meets an abrupt end in mid-curve.

(opposite) *Passage of a cosmic ray positron through a lead plate. An anti-electron enters bottom right at colossal speed and passes right through the lead plate, knocking out two other particles as it goes. A strong magnetic field applied to the cloud chamber curves its path to the left. This proves that although this anti-particle has the same mass as an ordinary electron it has the opposite electric charge*

All this was in complete agreement with Dirac's fantastic predictions.

The energetic ray had 'lifted the electron out of its hole' and the free hole moved like a positive electron until it met one of the many electrons present in all atoms and was obliterated! This *creation* of two oppositely charged particles is called 'pair production' and

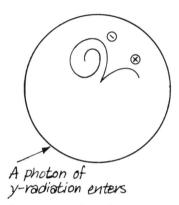

A photon of
γ-radiation enters

The creation of matter and anti-matter can also be seen on the last illustration in the book. Here minute hydrogen bubbles mark the trails of all charged particles and at the points B and C in the photograph two oppositely curving trails suddenly appear from nowhere!

showed the interdependence of matter and radiation in a new and striking way.

Since Dirac's theory applies to all matter, not just to electrons, one could expect anti-particles to exist corresponding to each and every particle known. However, for objects heavier than an electron so much more energy is required to form a pair that it would be a very rare occurrence indeed, even with cosmic rays. The anti-particle of the hydrogen nucleus—the anti-proton—was searched for in vain among cloud-chamber photographs and it was not until giant accelerators were built in the 1950s that the anti-proton was conclusively detected. It required nearly six billion electron-volts of energy to create a single proton–anti-proton pair and this was finally observed in 1955 by physicists of the University of California.

Now it became possible to imagine a whole world whose atoms were built up from this strange anti-matter existing somewhere in the universe like a mirror image of our own familiar world. This, too, had been forecast by Paul Dirac more than twenty years earlier. In his Nobel Prize lecture he had said:

> If we accept the view of complete symmetry between positive and negative electric charge so far as concerns the fundamental laws of nature, we must regard it rather as an accident that the earth (and presumably the whole solar system) contains a preponderance of negative electrons and positive protons. It is quite possible that for some stars it is the other way about, these stars being built up mainly of positrons (positive electrons) and negative protons. In fact, there may be half the stars of each kind. The two kinds of stars would both show exactly the same spectra, and there would be no way of distinguishing them by the present astronomical methods. (1933.)

The existence of such a star would present an appalling threat of total destruction to any 'normal' star near it in space. At first this idea seemed like the wild imaginings of the currently popular stories of Science Fiction, but when the new radio-telescopes of the 1960s located remote regions of space where 'quasi-stars' or quasars were generating incredibly large amounts of radio emission, it was seriously suggested that just such a pair of anti-matter stars or galaxies might be approaching each other and signalling to the universe their mutual end by this enormous burst of radio energy!

15
The Atom-splitting Years

Yet another result of Quantum Mechanics was a new approach to the nucleus of the atom. The pre-war model had contained both protons and electrons but by 1927 it became clear that the electron was too light a particle to be contained within the tiny nucleus.

In terms of Schrödinger's approach the size of the 'vibration pattern' of the electron is too large for the known diameter of an atomic nucleus. Alternatively by Heisenberg's Uncertainty Principle the precision involved in locating an electron in such a small space would allow such a large range of possible speeds and energies that it would escape too easily. It was true that many radioactive substances emit electrons from their nuclei but, as Schrödinger used to point out rather acidly, 'just because an electron is thrown out from the nucleus one cannot thereby prove that it had previously existed there in that form!'

The discovery of the Neutron

For some years Rutherford and others had suspected that another particle of *the same weight as the proton but electrically neutral* might exist within the nucleus. Long before there was any firm experimental backing for this new particle it had been christened the *neutron*. Finally, in 1932, James Chadwick, working under Rutherford at the Cavendish laboratory, proved that some very penetrating new rays produced by bombarding atoms with α-particles were simply a stream of the long-awaited neutrons. This new model of the nucleus, containing only protons and neutrons, is basically the one still used today.

The discovery of the neutron was to launch a new era in atomic

The New Nucleus
(1932 Style)

LITHIUM

⊕ - *proton*

● - *'neutron'*

Weight - 7 units
Charge +3 units

physics—a dramatic time of nuclear bombardment and atom-splitting—which was to culminate in the dreadful explosion of the first atomic bomb. In fact the first artificial disintegration of an atom had been effected by Rutherford as long ago as 1919. He had used the α-particles obtained naturally from radioactivity as 'bullets' with which to hit an atomic nucleus, but they were not very efficient. This was partly because of their relatively large weight and partly because they were not moving fast enough to overcome the electrical repulsion from the charged nucleus.

In 1928, before the neutron was discovered, J. D. S. Cockcroft (later Sir John) and E. T. S. Walton began designing an experiment in the Cavendish to break open the nuclei of some atoms in a much more controlled and deliberate way. They decided to use protons as their bombarding particles and, to make them more effective at penetrating through the electric repulsion and right into the heart of the nucleus, they accelerated their protons using a high electric potential (up to 600,000 volts).

This apparatus was novel and it was not until 1932 that they had their results ready to publish. The target used was made of the light metal lithium and the scintillations on the screen were shown to have been produced by α-particles (helium nuclei) although there was no helium in the apparatus at the start of the experiment.

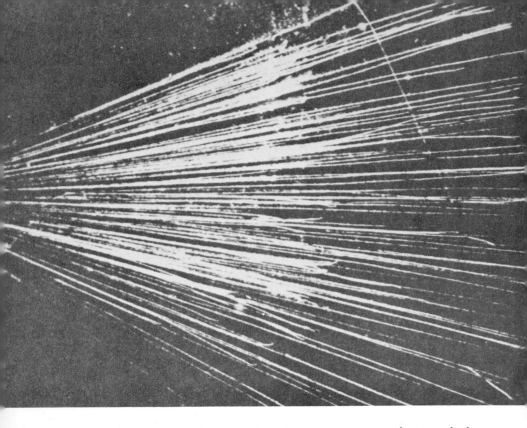

Rutherford's disintegration of a nitrogen atom, photographed by Lord Blackett. It was from this kind of photograph that Rutherford deduced that he had split an atom of nitrogen. The fast backward recoil of a particle from the collision breaks all the laws of billiards! It must have been furnished with extra energy by the atomic disintegration. Measurements proved that this was a hydrogen atom and that the particle recoiling in the other direction was oxygen

Cockcroft and Walton recorded the number of scintillations by observing the screen with a microscope. They estimated that for the best and thinnest target they could make, barely one proton in a million scored a direct hit on the nucleus of the lithium atom, was momentarily absorbed and then this new nucleus split into equal parts which were thrown apart with a great burst of energy.

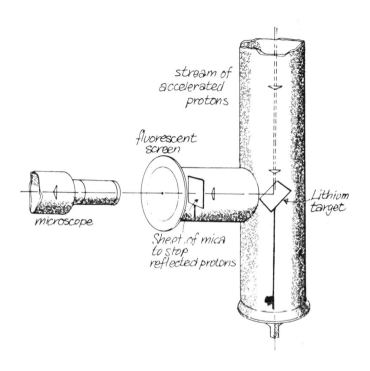

stream of
accelerated
protons

fluorescent
screen

Lithium
target

microscope

Sheet of mica
to stop
reflected protons

Cockcroft and Wilson's Apparatus

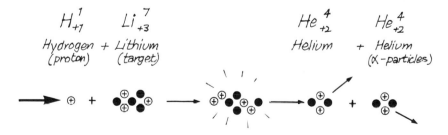

H^{1}_{+1} Li^{7}_{+3} He^{4}_{+2} He^{4}_{+2}

Hydrogen + Lithium   Helium   +   Helium
(proton)   (target)   (α-particles)

It was also possible to estimate the energy with which the two helium nuclei were thrown apart by interposing thin sheets of mica in their path to find out what thickness they could penetrate. It was immediately obvious that they had, between them, far more energy than the original accelerated proton which had started the whole reaction!

Mass and energy

This came as no surprise to Cockcroft and Walton. Though it is true that energy, like matter, cannot be created out of nothing the Special Theory of Relativity, by now some 28 years old, had shown that *mass itself was a form of energy*. Einstein had originally set out only to produce a logical system of space and time into which all the facts of experience and experiment would fit more convincingly; but the mathematics of the new Time-Space had thrown up the most famous equation of modern times:

$$ENERGY = MASS \times (Speed\ of\ light)^2$$

Einstein wrote 'Mass and energy are therefore essentially alike; they are only different expressions for the same thing. The mass of a body is not a constant; it varies with changes in its energy.' (*The meaning of Relativity*, 1922.) He had never asserted that mass would (or even could) be converted into energy in practice—merely that, looked at on the most basic level, they were of the same kind.

Earlier in the century it had been shown experimentally that a beam of very fast-moving electrons was harder to bend by a magnetic field than would have been expected. According to the Special Theory of Relativity this meant simply that the great energy of the electrons was reflected in their increased mass and accordingly had made them harder to deflect.

Later when 'pair production' was observed in cloud chamber photographs the creation of matter and anti-matter out of the energy of a cosmic ray could be seen as a more spectacular example of the same energy-into-mass conversion.

Now Cockcroft and Walton's figures could provide a numerical check with which to test Einstein's equation in a new way.

Mass before disintegration   1 proton = 1·0072 units
                                     1 lithium nucleus = 7·0134 units
                                     Total = 8·0206 units.
Mass after disintegration   2 helium nuclei = 2 × 4·0011
                                           = 8·0022 units

Loss in Mass = 0·0184 units

Using Einstein's equation this tiny loss should convert into

Gain in Energy = 17 million electron-volts

In fact the range of the α-particles indicated a liberation of energy of 17·2 million electron-volts which matched the predicted theoretical value very closely.

The World takes notice

On 2 May 1932 the news of this success was reported by the daily press which gave it dramatic coverage.

Never before had science so exploded into the headlines and caught the public eye. For months the home news had been full of the great depression, unemployment, and futile disarmament conferences. Abroad the Sino-Japanese War had just broken out and in Germany the Nazi party was rising to power amid public invective and mob-violence. To some extent, therefore, the newspapers seized upon Cockcroft and Walton's experiment as an escape from drearier topics into a world of 'alchemical' magic, 'weird scientific machines', and patriotic pride. Despite its quotation of Cockcroft's own words 'immense scientific, but no practical importance' the *Daily Mirror* tried to lend more drama to the story with the blood-chilling comment

. . . We shudder as we hear about the disintegrations . . . that may blow our planet to pieces . . . some new horror for the League of Nations to forbid (vainly) in time of war.

Nevertheless the significance, in that time of economic depression, of the new form of energy was clearly seen. Several daily papers underlined this promise of more power for industry. The *News Chronicle* even gave it a front-page headline:

'Man's control of ENERGY',

and the *Daily Express* commented soberly—'The next task of science is to make use of the energy liberated by this disruption of matter. At present only one in 100,000,000 electrical particles manages to split the atom. The coaxing of the rest has to be attempted.' But it was across the Atlantic, in the *New York Times*, that the news was given its most impressive treatment. Next to, but in bigger type than, F. D. Roosevelt's plans for nomination as presidential candidate, the headlines read:

'ATOM TORN APART YIELDING 60% MORE ENERGY THAN USED.'

The following day the idea of energy from the atom was still exciting headline interest and, on the third day, the *New York Times* printed a thoughtful editorial on the subject entitled, simply but prophetically, 'Atomic Energy'.

In some ways the reporters were actually ahead of the scientists. In vain Rutherford and Cockcroft emphasised the purely theoretical interest of their experiments; the public had a nose for economic progress. Only a month before Aldous Huxley had published *Brave New World* and other science-fiction stories abounded so that startling wealth and progress was almost daily expected from science. It was to be thirty years yet before the first nuclear power station went into commercial production but the reporters could feel it on its way. It was the physicists, wrapped up in their interest in the structure of the atom, who were blind to the economic revolution that they might cause. They had, as it turned out, only seven years more of 'pure' atomic research ahead of them before the pressures of economics and war forced them into government committees and research departments where the luxury of intellectual curiosity was to be replaced by supervised war effort.

The nuclear laboratories get busy

Those seven years were spent largely in devising more ambitious atom-splitting experiments although some progress was also made in understanding the structure of the nucleus. In the same year that the neutron was identified Werner Heisenberg published the first paper on modern nuclear theory, and he was shortly to be followed by Niels Bohr and the eminent Italian physicist, Enrico Fermi. Two general points were stressed in all these early attempts to probe into the tiny heart of the atom—composed, it was now accepted, of a mixture of neutrons and protons.

1 A new, unknown force of enormous strength must bind together the neutrons and protons far tighter than gravitational, electrostatic, or magnetic attraction. This new force was appropriately named the *Strong Force* but little was known about it either then or, indeed, now. (The normal electrostatic *repulsion* between the charged protons in the nucleus must clearly be much weaker for all stable atoms.)

2 Every atomic nucleus contains great 'bonding energy' since the total nuclear mass is always slightly less than what would be obtained by adding up the masses of all the protons and neutrons it contains. This lack of weight, or 'mass defect', as it is called, varies from atom to atom but from it can be calculated the energy of formation of the nucleus using Einstein's equation once more:

$$\text{Bonding } Energy = Mass \text{ Defect} \times (\text{Speed of light})^2$$

Pictorially this means that if the appropriate number of free protons and neutrons could be assembled near enough to each other, the new Strong Force would hurl them together with such a colossal impact that enormous energy would be liberated and the corresponding amount of mass would be lost!

The Strong Force only operates at extremely close range so that it is beyond our powers to press together neutrons and protons in order to build directly from them a new atomic nucleus. It may be

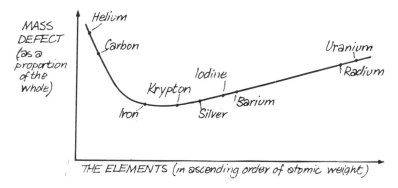

THE ELEMENTS (in ascending order of atomic weight)

that such a process does take place within the interiors of massive super stars and generates the energy to keep them at 'white heat'.

The graph shows clearly that it is the nuclei of middle atomic weight which are most tightly bound together and so, if they could be formed out of either much heavier or much lighter atoms, the largest amounts of energy would be liberated. However the experimental physicists of the 1930s were still not primarily involved in a search for available energy. The 'atom-bashing' experiments of the next six years were based on the idea of either chipping a little bit off the nucleus, or adding a little bit on.

Atoms of every sort were being bombarded with neutrons or other particles in France, Italy, Britain, Germany, and America and much of interest was learnt. When ordinary atoms captured a neutron they often became unstable and radioactive themselves. These 'radioactive isotopes' are now widely used as 'tracer elements' in medicine, agriculture, and industry since their progress through any system can be recorded by the rays they continuously emit. Fermi, with his newly recruited Italian team of atomic physicists, was the first to bombard the heavy uranium atom with neutrons. This should have been the most dramatic occasion but something went wrong with the experiment. It is difficult to see now where the fault lay—he detected various types of radiation, it is true, but nothing outstandingly energetic. Since the activity was of a new period of decay and as he could not identify the chemical nature of the radiating atoms so formed he assumed, tentatively, that he had made a new element heavier by one or two units than the heaviest element then known—uranium itself.

Neither of these elements is found naturally but they have now been identified definitely and named *neptunium* and *plutonium*. Though the electrons could be detected Fermi was laudably cautious in the conclusions that he drew:

> . . . suggest the possibility that the atomic number of the element may be greater than 92. . . . This hypothesis is supported to some extent. . . . However this evidence cannot be considered as very strong. (*Nature*, 1934.)

No such caution troubled the fascist press of Italy! Breathing down the necks of scientists had now become an international pastime for newspapers and the 'new element' was widely reported under headlines such as 'Fascist Victories in the World of Culture'. Fermi was unsure and distressed by this premature publicity, but little did either he, or the fanatic followers of Mussolini, realise how much greater a discovery had been missed!

16
War and Fission

December 1938 was perhaps the dawn of the 'Nuclear Age'. The very title is a little fearsome to us now and indeed the times themselves were grim. Hitler's persecution of the Jews uprooted many famous men in the world of science. Einstein and Fermi (whose wife was Jewish) emigrated to the United States, Max Born had already settled in England, and Lise Meitner, head of Physics in the Kaiser-Wilhelm Institute of Berlin, fled to Stockholm—not to mention scores of other lesser-known scientists. Max Planck tried to intercede with Hitler for his colleagues but was only screamed at for his pains, while the eccentric but upright Erwin Schrödinger went into voluntary exile in protest against this persecution.

In Paris the Joliot-Curie's (daughter and son-in-law of Marie Curie) had again been bombarding uranium with neutrons and thought that they could detect a rare middle-weight element, lanthanum, but it seemed unbelievable. In Berlin Hahn and Strassman to their astonishment found another middle-weight element in their uranium residues, this time barium. It began to seem as if the heavy uranium nucleus had actually split in half—this was quite unexpected and completely different from any of the 'atom chipping' experiments of previous years. The discoverers, who were chemists themselves, felt uneasy and wrote, 'We cannot decide to take this step in contradiction to all previous experience in nuclear physics'. The physicists were bolder. In Sweden Lise Meitner and her nephew Otto Frisch together worked out the implications of this novel type of nuclear reaction and published their results in January 1939. Frisch has said of his theory:

It looked as if the absorption of the neutron had disturbed the delicate balance between the forces of attraction and the forces of repulsion inside the nucleus. It was as if the nucleus had first become elongated and then developed a waist before dividing into two more or less equal parts in just the same way that a living cell divides.

Because of its similarity to cell division and reproduction, Frisch named this process *nuclear fission*.

This borrowing of a term from biology was a small reminder of the new synthesis of the natural sciences that was already on its way. In the 1930s the first experiments had been performed to show the

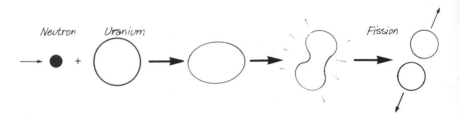

Neutron   Uranium   Fission

genetic effects of X-rays on cells—the 'atoms of life'. The rate of mutation (permanent and inheritable changes in the cell) had even been compared quantitatively with the energy of the radiation causing it so that, while atomic nuclear fission was pursued for the energy it might give, the dreadful effects of this within the biological nucleus of the human cell were already predictable, years before the bomb was dropped on Hiroshima.

If such fission had indeed taken place an astounding quantity of energy would be liberated—in the region of 200,000,000 electron-volts from a single atom. This is enough to make a heavy grain of sand jump visibly! Weight for weight it is three million times the energy of burning coal and twenty million times the energy of exploding TNT. The source of this vast energy is indicated on the mass defect graph; if the over-heavy uranium atom splits into a couple of middle-weight ones there will be a large loss of mass which must reappear as a huge burst of energetic radiation. Immediately Frisch in Copenhagen and Joliot-Curie in Paris demonstrated this production of energy experimentally and, when Meitner and Frisch's paper was published, physicists all over Europe and America rushed into their laboratories to confirm the result. It was really astonishing that it had not been detected before. There is an amusing story told of a Cambridge physicist who, when bombarding uranium with neutrons some years earlier, found that his home-made Geiger-counter had recorded occasional pulses of enormous energy. Assuming that his instrument was misbehaving he 'fixed' it so that such unbelievable readings could not occur again—and then continued with his routine experiment!

The vibrating drop

Niels Bohr, now 54 years old, was once more in the middle of the latest atomic developments. This was partly because Frisch had emigrated to Copenhagen to work with him, and partly because he was visiting the States and in contact with Fermi when the news of nuclear fission broke. As always his fertile, pictorial imagination was stimulated and he evolved a model of the nucleus which helped both to visualise and calculate the processes of fission. This is known as the 'Liquid-drop' theory. In many ways the behaviour of the nucleus of an atom is like that of a drop of liquid whose agitated molecules are bound together by the forces of surface tension. Large drops, like large atomic nuclei, are less stable than small ones and, given energy, tend to reduce their size either by evaporation or by splitting in two.

This model—so simple to visualise and understand—does in fact describe both the ways that an actual uranium nucleus can react to the impact of a neutron. As Bohr himself wrote:

> In the case of ordinary reactions, resulting in the emission of a proton, neutron or α-particle from this nucleus, we have to do

with a concentration of a considerable part of the excitation energy on some particle at the nuclear surface, sufficient for its escape, which resembles the evaporation of a molecule from a liquid drop. In the case of the fission phenomena, the energy has to be largely converted into some special type of motion of the whole nucleus causing a deformation of the nuclear surface sufficiently large to lead to a rupture of the nucleus comparable to the division of a liquid drop into two droplets. (*Physical Review*, 1939.)

In the same paper Bohr calculated that whereas the commonest form of uranium (atomic weight 238) would be most likely to react in the first way—that is to absorb a neutron and then emit one or two particles from its surface—the uranium nucleus most likely to

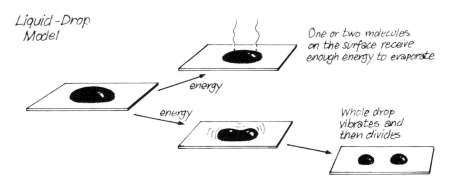

Liquid-Drop Model

energy

energy

One or two molecules on the surface receive enough energy to evaporate

Whole drop vibrates and then divides

undergo fission is that of the lighter isotope (atomic weight 235). Both these forms of uranium occur naturally but uranium-235 is much rarer, being only about one part in 140 of the normal metal.

Chain-reaction

In the case of an elongated, vibrating, drop of liquid, when the final moment of disruption occurs the two droplets produced will not necessarily be of exactly the same size. Chance, it seems, will play its part. So also in nuclear fission a wide variety of roughly middle-weight atoms have been identified among the residues.

Barium, krypton, strontium, cobalt, iodine, silver, and bromine are just a few of the atomic fragments known to be formed. Many scientists were quick to realise that the pairs of elements produced all had fewer neutrons between them than the original uranium nucleus. This meant that every time a nucleus was split by a single neutron it would emit several more energetic neutrons—each one of which might start another nuclear disintegration, and so on. In this way a 'chain reaction' could be started with one neutron which might spread like a bushfire throughout the whole mass of uranium. Here, at last, was a feasible source of great power.

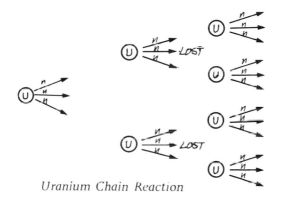

Uranium Chain Reaction

Joliot-Curie, in Paris, was one of the first to confirm this generation of extra neutrons. He estimated that between three and four more neutrons were liberated for each nucleus that was split. It seems that this figure was too optimistic and later Fermi showed that the average number produced was about 2·5—a closely guarded wartime secret. The French also knew that these new swift 'fission neutrons' would have to be slowed down in order to cause more nuclear disintegrations and that either heavy water or graphite might bring this about. With astonishing commercial flair they patented a design for the first ever chain-reacting atomic pile at a speed that left other scientists gasping somewhat resentfully!

Hitler on the march

The attitude of the scientific world may have been caused by the tough competitive approach of the French, or because the whole design of their pile was based on little more than inspired guess-work; but little inspiration was needed at that time to foresee the outbreak of another violent World War. British and European scientists sensed at once the dreadful possibilities of the atomic fission chain-reaction in this context. Gone were the 'ivory-tower' scientists of 1932—in their place were desperate men pleading for money from their governments and haunted by the thought of what their erstwhile colleagues across the enemy lines might have already begun to make. In March 1939 Hitler's army occupied Czecho-slovakia and seized with it the richest uranium mines in Europe. Within days of this sinister event bewildered military men in Britain and America were being treated to earnest lectures on nuclear fission by patriotic professors who urged them to devote immense resources to develop a new and incomprehensible explosive. In Great Britain the threat of war was close and the agitating professors were known and eminent men like James Chadwick and G. P. Thomson, so the government listened and acted on the advice given. In America it was the distinguished refugee physicists like Fermi who, knowing the evils of the Nazi movement at first hand, tried in their broken English to warn the complacent men in power. Perhaps it is little wonder that they were less successful there than on the other side of the Atlantic.

In Britain G. P. Thomson approached the Air Ministry with a specific request for one ton of the purest uranium oxide to be found for research into its military potential. He was given the appropriate funds and returned to Imperial College, in his own words, 'like a character in a third-rate thriller'. Here he carried out some very basic experiments and found, at once, the two major difficulties that scientists were going to have to solve in the frantic struggle against time which lay ahead. The hoped-for chain-reaction would *not* take place within his sample of uranium because the new generation of fast neutrons liberated by fission were largely swallowed up by the more common isotope, U-238. There were two theoretical solutions:

1   To separate the rare fissionable U-235 from the other isotope. Since these are chemically identical Thomson dismissed the idea at that time as impossible.

2   To slow down the issuing neutrons to about 1/1,000 of their initial speed so that they would be more likely to cause fission in U-235 than to be uselessly absorbed in U-238. To do this the neutrons had to collide many times with light, non-reacting, atoms. The double-weight isotope of hydrogen which makes up 'heavy water' was very effective but too rare, Thomson thought, to be available in the quantities that would be needed.

Sometime in early 1940 the British team of nuclear scientists started making headway. Frisch, R. E. Peierls, and Chadwick pooled their ideas on how to separate U-235 from U-238 and evolved a method based on the very small differences in density of the vapours made from these two isotopes. In March, they set up a committee known familiarly as 'Maud' to co-ordinate research into the chemical and physical problems that had to be tackled.

In April of the same year Norway was occupied by German forces. This was another event to alarm scientific circles since the only factory in the world which was equipped to manufacture the 'heavy water' so vital for continuous nuclear fission was situated in Vemork, southern Norway. Then, as soon as the plant was in German hands, a sinister order was issued to increase its production to 3,000lb of heavy water per year!

The first atomic pile

Once more Fermi applied for research funds and this time he received a meagre $6,000 from the American Army and Navy Departments. He spent this money on the purest uranium oxide and graphite that he could buy and began the construction of what was to be the world's first nuclear reactor. The graphite, pure carbon, was used in place of heavy water, to slow down the neutrons emitted by fission. The University of Chicago provided a disused squash-court and eventually on 2 December 1942 Fermi kindled the first 'atomic fire' and kept it under control. The steady clicking of radiation counters was the only indication that the atomic pile had 'gone hot'. It was allowed to 'burn' for twenty-eight jubilant minutes before it was 'put out' by inserting neutron-absorbing rods into the heart of the reactor. This success was then cabled to the White House in the famous coded message—'The Italian navigator has reached land and found the natives friendly'.

17
The Bomb

But what of the bomb? Looking back now at those early years of war when scientists first began to plan and make this modern symbol of human catastrophe it is impossible not to feel an overwhelming urge to turn back the clock and shout out a desperate warning to stop before it was too late! What sort of power-mad men were they? Had they no conscience or imagination? The answer is disarming. Although brilliant and inventive, these were mostly reluctant men and, of the prime movers, three at least were uncommonly gentle and horrified by violence. Up until the outbreak of war science had known no national frontiers and Hitler's persecution had at first actually enhanced the exchange of ideas as scientists fled from one country to another. In Britain and America the leading scientists recognised only too well the worth of German physics which had led the world for two decades. Many of them had known Heisenberg, Hahn, and others personally and admired them. When rumours came out of Germany of new laboratories dedicated to research into uranium Allied physicists shuddered at the possibilities. Right until 1944 they believed that the Germans were actually ahead of them in bomb production and that this was the truth behind Hitler's boast of a terrible 'secret weapon'. The reality was very different but they could not have known. While they struggled with the technical problems of making this devastating weapon it was the fear that Germany would be first and would drop it on their own homes and cities that drove them on. Such fear can very easily drive out caution about the possible results of such destructive power in the hands of any nation. It was only the blinding flash of the first explosion, when the work was done, that seared their minds with doubt.

In America there were always refugee physicists who could foresee the terrible destruction that fission might bring about. Leo Szilard, a Hungarian, was the first to urge contact with the American government, yet he was also the first to campaign against dropping the finished bomb on Japan. In 1939 it was probably both the fear of Germany and the insecurity of exile that made him draft a famous letter to President Roosevelt. He and his friends believed that the only scientist to whom the president would listen with respect was Albert Einstein. This was a man who had been a life-long pacifist, humanitarian, gentle and remote, yet such were the times that after little persuasion Einstein agreed to sign this prophetic letter.

F. D. Roosevelt
President of the United States
White House
Washington D.C.   Aug. 2nd 1939
Sir,

Some recent work by E. Fermi and L. Szilard, which has been communicated to me in manuscript, leads me to expect that the element uranium may be turned into a new and important source of energy in the immediate future. Certain aspects of the situation which has arisen seem to call for watchfulness and, if necessary, quick action on the part of the Administration. I believe therefore that it is my duty to bring to your attention the following facts and recommendations.

This new phenomenon would also lead to the construction of bombs, and it is conceivable—though much less certain—that extremely powerful bombs of a new type may thus be constructed. A single bomb of this type, carried by boat and exploded in a port, might very well destroy the whole port together with some of the surrounding territory. However such bombs might very well prove to be too heavy for transportation by air.

I understand that Germany has actually stopped the sale of uranium from the Czechoslovakian mines which she has taken over.

That she should have taken such early action might perhaps be understood on the ground that the son of the German Under-Secretary of State, von Weisaker, is attached to the Kaiser-Wilhelm Institute in Berlin where some of the American work on uranium is now being repeated.

Yours very truly,
A. Einstein.

Roosevelt read the letter, listened to the arguments, and exclaimed in sympathy, 'What you are after is to see *they* don't blow *us* up'. It was in this defensive spirit that the project of the bomb was conceived. However America was still two years away from direct involvement in the war so, at first, interest was sluggish and progress slow.

The next year a discovery was made, foreshadowed years before by Fermi, that was to have tremendous implications for bomb manufacture. The object was an innocent-looking smear of uranium on a cigarette paper which had been irradiated with neutrons. Apart from fission in the occasional atom of U-235 it was found that the common isotope U-238 had absorbed a neutron and changed first into one new element and then into another. These were named after the planets beyond Uranus in the solar system.

$$\text{Uranium-238} + \text{neutrons} \rightarrow \text{Neptunium-239} \rightarrow \text{Plutonium-239}$$

The first, neptunium, was unstable and changed quickly into the second element, plutonium. Never was an infernal substance so well named! It was recognised at once as being fissionable like U-235 but with such a blaze of radiated neutrons that two small pieces could not be brought close together without exploding with immense energy. Not all of this was yet known for certain though the potentiality was clear, but such was the remoteness of America from war in the summer of 1940, that the discovery was published for all the world to see in the pages of the *Physical Review*!

These were terrible times for Britain. Her scientific manpower was stretched to the full by the effects of war and the continual need to develop radar against enemy bombers. British nuclear technologists were actually ahead of America at this moment but they were more than willing to share all their secret plans for separating U-235 from natural uranium if only America would relieve them of the burden of production. They shocked their trans-atlantic colleagues by the blunt way they used the word 'bomb', they preached explosive fission not controlled fission reactors. Roosevelt wrote to Churchill suggesting co-ordination. In December 1941 the decision was finally taken to go all-out for the production of an American nuclear bomb. The very next day Pearl Harbour was attacked by the Japanese and the United States entered the war.

Leading American scientists began at once to plan methods of producing the two known fissionable elements—U-235 and pluto-nium. In addition they needed an exceptional theoretical physicist to examine the problems of making and detonating such a bomb. Fortunately such an outstanding man stood ready for the challenge, a native-born American moved to uncharacteristic hatred by all that Nazi Germany stood for.

Robert T. Oppenheimer was a creedless Jew, a tolerant and cultured man with a brilliantly clear understanding of modern physics. He had worked on Quantum Mechanics with Dirac, on cosmic rays, and on the theory of exploding and collapsing stars; he was an inspiring teacher and enjoyed an international reputation. Although later in the 'witch-hunting' period of the McCarthy era he was accused of being a communist, the philosophy which most gripped him was that of India. He saw in 'Ahimsa'—harmlessness or non-violence—a deeply sympathetic doctrine which remained with him all his life. This then was the man who, in greatest measure, brought about the most dreadful explosion known to man!

Oppenheimer's first task was to calculate just how intense a heat this explosion would produce. In June 1942 he summoned a conference of seven theoretical physicists to study the problems. They met in a locked room behind barred windows in Berkeley, California, and the results of their calculations were so terrifying that Oppenheimer suspended the conference at one point while he decided whether or not to abandon the whole project. They had computed a temperature, produced by a fission bomb, so high that it might cause the heavy hydrogen in the sea to fuse into helium or the nitrogen in the air to transmute into carbon. Both these reactions would liberate, in turn, huge quantities of energy as Oppenheimer knew only too well, for they form the basis of the interior heat of all stars and of the sun itself. The eight men stared at their calculations in horror! Would the first nuclear bomb ignite the very oceans and atmosphere of our planet in a final annihilating blaze? As the eminent scientist Arthur Compton remarked at the time, 'Better to be a slave under the Nazi heel than to draw down the final curtain on humanity.' A week later when the conference was reconvened they found a small but reassuring mistake in the calculations. Tensely and quietly they reassessed the chances of total calamity and found them very small. Although after this the production of the bomb went ahead there were always physicists, right until the moment of the final test, who still harboured in their minds this frightful uncertainty.

Shortly after this the Army took over the whole complex project. Two new cities had to be built with giant factories to produce the raw materials of the bomb. Oppenheimer's section which was to make the bomb itself in utter secrecy also needed a remote building site for a new town and this was swiftly found. Years before when he was recovering from tuberculosis, Oppenheimer had bought himself a ranch high up in New Mexico. The austere magnificence of the desert moved him as it has always moved deeply feeling men so that he knew at once that the two loves of his life would always be physics and the desert. Now he had a chance to combine them both. In a wild isolated place where cottonwoods bordered on the

Alamagordo, New Mexico . . . Trinity Test, 16 July 1945.
Ground Zero is marked by a barren mound of earth and a
weathered stake

reddish hills and canyons of the desert he founded the secret township of Los Alamos where he was to live and work for the next four years.

A group of top-ranking scientists was quickly recruited and worked in this lonely but exotic landscape with a kind of intellectual excitement and dedication that only Oppenheimer could have inspired. Discussion was free and Oppenheimer listened to every suggestion with both courtesy and penetrating comprehension. Two different types of bomb had to be designed for uranium and for plutonium. Each involved fascinating problems in totally new fields but every difficulty was inventively overcome. Through it all Oppenheimer found the magnificent presence of the surrounding desert an inspiration to his belief in Ahisma as well as to theoretic physics. And when the long day was over he read himself to sleep with the 'Holy Sonnets' of John Donne. The times, the task, and the men were all beyond the common run and from them sprang a bizarre mixture of poetry, physics, and destruction during those early years at Los Alamos.

By July 1945 one uranium bomb had been assembled and an intricate device for exploding the more reactive plutonium had been designed. This was so complicated that a test of the bomb was essential and it stood ready on the night of 14 July fixed to the top of a tower and illuminated by occasional flashes of prophetic lightning. At 5.30 the following morning it was exploded.

Groups of the physicists stood and watched, some ten miles away, some twenty, while the huge fireball consumed the desert night. It has been described as an unearthly green and as a searing, blinding white. For the first time since the world began it was flooded with a dazzling brilliance beside which the coming dawn paled into insignificance. Awful and majestic the column of quivering cloud rose higher than any mountain on earth. Then, through the spellbound silence, came the mighty blast of heat and sound that reverberated until the ground trembled a hundred miles away. Prostrated, literally, on the ground, the violence of their own creation moved

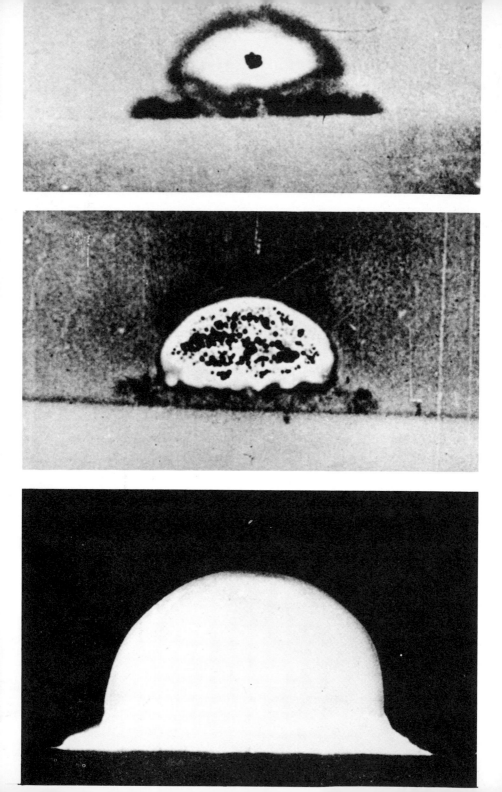

the watchers personally in a way they had never imagined. Some of them wept, some laughed, and for several days the sight of such threatening power haunted them continuously. In the blinding glare of actual fission they understood in a new way the reality of the primordial energy, locked within the nuclei of atoms, which they had let loose on earth.

In Oppenheimer's mind the sight evoked a passage from the sacred Gita of the Hindus—'I am become Death, the Shatterer of Worlds'. Was Krishna, who is also compassion and creation, in that awesome cloud? Was this the force that would end war? So charged with emotional power was this first, ominous 'happening' that it was to reach across the intervening years and mould a generation of gentle 'Hippies' who, like Oppenheimer, studied Indian philosophy and longed for peace. But a little distance away his second in command was weeping like a child, 'they are going to take this thing and fry hundreds of Japanese', he cried.

Many of the scientists were against dropping the bomb on Japan. Some protested, but it did no good. The military convinced President Truman that it would cost the lives of a million American soldiers to end the war in any other way and he took the final decision. On 6 August the uranium bomb was dropped on Hiroshima seventy thousand people died, and on 9 August the plutonium bomb was dropped on Nagasaki, killing another thirty-five thousand. No one at Los Alamos cheered.

After that the war ended quickly but the mood at Los Alamos had changed. When the Army publicly commended the laboratory Oppenheimer replied in a bitter speech,

(opposite) *Initial test of the atomic bomb in New Mexico on 16 July from a distance of 6 miles.* Above: *The start of the explosion. This small cloud later rose to a height of 40,000 feet.* Centre: *Multi-coloured cloud from the explosion. Black areas were brighter than the sun itself, according to observers.* Below: *A later stage of the development of the cloud*

. . . the time will come when mankind will curse the names of Los Alamos and Hiroshima.

Some of the original team stayed to work on further bombs but even the government lacked enthusiasm at first. America believed that it was twenty years ahead of Russia or any other nation and was concerned only with using this imagined advantage to secure peace and the international control of atomic weapons. But one of the trusted British physicists at Los Alamos had been Karl Fuchs, a Russian spy, and by 1949 America was awakened by the detonation of a Russian atomic bomb. To recapture the lead they could only commit themselves to the illusive 'super' bomb that Edward Teller and others had never forgotten. A plutonium fission bomb should be hot enough to cause the double or treble weight isotopes of hydrogen to fuse into heavier atoms. This had always been a possible line of development on which Teller and Fermi had speculated as early as 1943.

$$_{+1}^{2}H \; + \; _{+1}^{2}H \; \longrightarrow \; _{+2}^{3}He \; + \; n \; \longrightarrow$$

$$_{+1}^{2}H \; + \; _{+1}^{2}H \; \longrightarrow \; _{+1}^{3}H \; + \; proton \longrightarrow$$

$$_{+1}^{2}H \; + \; _{+1}^{3}H \; \longrightarrow \; _{+2}^{4}He \; + \; n \; \longrightarrow$$

The loss of mass in such nuclear reactions is relatively much greater than in fission and in 1950 the government gave the 'go-ahead' to Los Alamos to make the hydrogen bomb in spite of the reluctance of some respected scientists to create yet more destructive power.

For Oppenheimer and others, however, the turning-point had come. Neither simple patriotism nor the mysteries of the atomic nucleus could hold them at Los Alamos after the war. They had seen the results of their labours in terms of human deaths and mutilations. The furious heat of 'their' bomb had vaporised completely one whole human being, the shadow of whose body is still preserved in the concrete at Hiroshima, and had maimed within the ovaries of young girls a generation yet unborn. The conscience of the world was appalled. In India Mahatma Gandhi, who had used Non-Violence as a political force to free his country from British rule, wrote of the moral dangers involved in even planning such a bomb. On the very day that he was assassinated Gandhi had taught his followers the only thing that this terrible bomb could never destroy is the power of Ahimsa itself! Oppenheimer's gentle soul was sickened by what he had done and he said publicly 'The physicists have known sin'. He went from one university to another trying to recapture his peace of mind in the serenity of cosmic ray research, but his telephone rang continuously. The President planned and replanned his atomic strategy and Oppenheimer was summoned to act as adviser on every Atomic Commission for the next six years—a saddened and reluctant expert. The last indignity was an examination of his politics and loyalty in which he was publicly branded as a security risk. He was allowed no further access to his country's atomic secrets and returned to the seclusion of an academic life and the physics of the new elementary particles.

18
New Pieces of Matter

In the early 1930s the list of simple particles out of which all matter is built had seemed to be satisfyingly complete. The two charged particles of Rutherford's first atomic model were:

1 *The Electron*   Negative and light-weight
2 *The Proton*   Positive and about 2,000 times
  (Hydrogen nucleus)   heavier

To these the quantum theory had added:

3 *The Photon* (or quantum) A quantity of pure electromagnetic energy travelling at the speed of light
4 *The Positron*   Positive and light-weight
  (anti-electron)
5 *The Anti-Proton*   Negative and about 2,000 times heavier

The last building brick of ordinary atomic nuclei was discovered in 1932:

6 *The Neutron*   Neutral and slightly heavier than the proton

Theory insisted that there should also be an anti-particle to this which was only identified years later:

7 *The Anti-Neutron*

This scheme seems simpler if the elementary particles are seen as the constituents of matter or anti-matter, the self-destroying antithesis of each other.

But in science we are fated never to rest on our laurels and admire

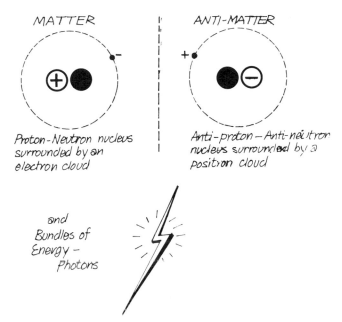

MATTER | ANTI-MATTER

Proton-Neutron nucleus surrounded by an electron cloud

Anti-proton—Anti-neutron nucleus surrounded by a positron cloud

and
Bundles of
Energy –
Photons

the 'final simplicity' of our solution for long! Nature always tempts the inquisitive minds of the next generation to upset the old order and to construct a new one. So by 1936 a search for fundamental particles had begun again.

The Mesons

The mysterious 'strong force' which holds the protons and neutrons together in the nuclei of atoms was quite new to physics. It was clearly not electromagnetic in origin since it binds one charged particle to another similarly charged as well as the uncharged neutrons. It was known to be both hundreds of times stronger than any other interaction between matter and very short range indeed, being imperceptible at distances greater than a millionth of a millionth of a centimetre! In 1935 a Japanese physicist, Yukawa, examined this problem in the manner of quantum mechanics. The energy of electrical attraction between oppositely charged particles can be released as photons or quanta. The energy of the strong nuclear force is very different but could also be expressed as quanta,

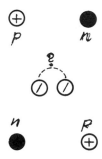

Interchange of a 'Strong force' quantum changing a proton into a neutron and binding them together

of a sort, which would reside within the nucleus binding the protons and neutrons together as it was tossed from one particle to the other. Yukawa's mathematical analysis showed that these 'quanta of the strong force', if set free, would not be weightless, like photons, but would have a mass about 300 times that of the electron. Did they ever escape from the nucleus? Yukawa predicted that they would have charge as well as weight and wrote,

> Such quanta, if they ever exist and approach matter close enough to be absorbed, will deliver their charge and energy to the latter. . . . The massive quanta may also have some bearing on the shower produced by cosmic rays. (*Progress of Theoretical Physics*, 1935.)

The challenge to experiment was immediately taken up. Because their weight was in between that of electrons and protons they were called 'mesons' and looked for, as Yukawa had suggested, in the streams of particles produced by cosmic radiation. Only the protons which tear into our upper atmosphere at colossal speeds from outer space could have the energy to release these mesons by violent collision with a nucleus. Between 1936 and 1938 cloud-chamber photographs had demonstrated tracks of just such intermediate particles and Yukawa's theory seemed to be triumphantly vindicated.

In reality matters had not progressed quite so smoothly. The new particle —the μ-meson as it was called—was shown to be a little too light to fit Yukawa's theory and, what was much worse, it did not have the energetic reaction with nuclei that such a 'quantum of the strong force' should have. On the contrary it could, on occasions, disappear into an atom quite peacefully taking the place of an ordinary bound electron. Not only had Yukawa's meson not yet been found but the new particle seemed a quite unnecessary addition to the existing scheme! It was a pseudo-electron about 200 times as heavy as the familiar variety and could be either positive, negative, or neutral.

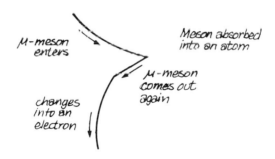

After the war, in 1947, another attempt was made to find Yukawa's meson. This time photographic emulsion was used to record the traces of charged particles and these plates were exposed to cosmic rays high up on the Pic du Midi at an altitude of about 9,000 ft. At last traces of a particle with the right characteristics were found and it was named the π-meson. The reason why they had so long eluded detection was easily explained for, though produced plentifully in the upper atmosphere, they are very short-lived and rarely reach sea-level. Further experiments showed that they reacted with atomic nuclei with the right amount of drama. Being bundles of strong-force energy they fling out particles in all directions and such 'events' have now been recorded many times.

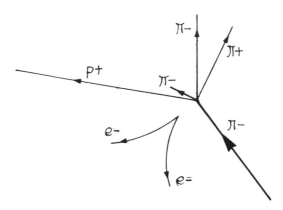

Collision of a π-meson with a proton giving rise to:

> 2 negative π-mesons
> 1 positive π-meson
> γ-radiation (photons) one of which later creates 'a pair' of electrons
> flinging out the original proton.

Now the number of the so-called 'elementary' particles had been increased by four μ-mesons and three π-mesons making a total of at least fourteen—and worse was to follow. By examining rare events of high energy in cosmic radiation more and more new particles began to appear. In 1953, when the first giant accelerator of protons came into research use these occurrences became more common, particles both lighter and heavier than the proton were produced and so curious were their properties that they were called 'strange particles'. By 1964 the number of different particles known to exist in nature rose to a horrifying and unwieldy total of over 80! Where

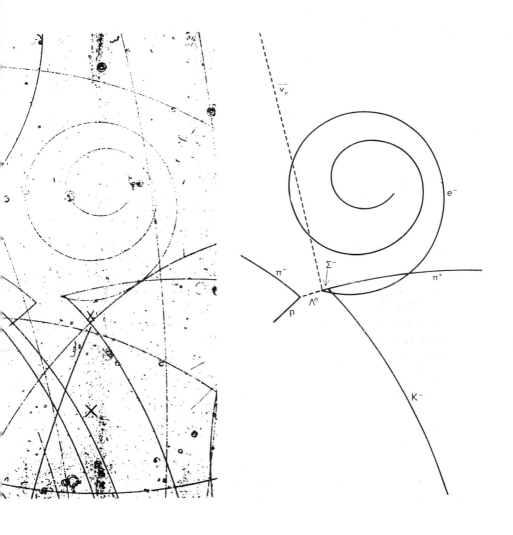

'One of the strange new particles Sigma Minus (Σ^-) which was discovered in a bubble chamber photograph. From the length of its track it can be shown to have existed for less than a millionth of a millionth of a second!'

now was the magnificent simplicity that science had striven towards for so long?

The search for rules of symmetry

The last two decades have been a bewildering period in particle physics—perhaps it will prove to be comparable to the similar span of time fifty years ago when the Quantum Theory was just emerging. Then physicists had had to adjust to the idea of a new symmetry between particles and waves which paved the way to a better understanding of matter and light. Now we have another problem. Can we incorporate this new multitude of basic particles into a scheme which will make sense of their various tiny masses and rapid disintegrations? Is there any enduring pattern beneath the sudden births and deaths of these fugitive units of matter?

Time and change are the essence of physics and the very instability of the new particles was a challenge to examine anew the symmetry of inter-changes in space. Some of the 'strange' particles exist for no more than a millionth of a millionth of a second, some for even less, but the traces of these transformations can give a lot of information if they follow the patterns of established laws. For instance the sum total of the Mass-and-Energy during any change must remain the same—this is a symmetry with respect to time which is the modern form of the Law of Conservation of Energy. In the 1930s Fermi had discovered an apparent leakage in this law in the field of radioactivity. He postulated the existence of a neutral particle, perhaps similar to the electron, which robbed the system of its energy, and named it the *neutrino*, 'little neutral one'. Angular rotation, spin, and momentum must also be conserved and when losses in all these quantities were detected that, too, was laid at the door of the neutrino.

The most obvious symmetry that physicists have always assumed is that of *mirror-reflection* and they called it the Conservation of Parity. Right-handed and left-handed objects and rotations do exist in the living world and have always had a deep significance in

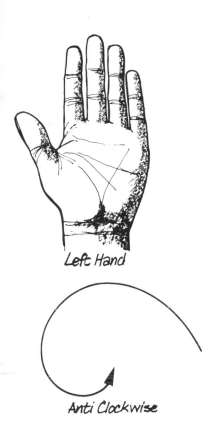

Left Hand

Anti Clockwise

l-Glucose

Twists polarised Light to Right

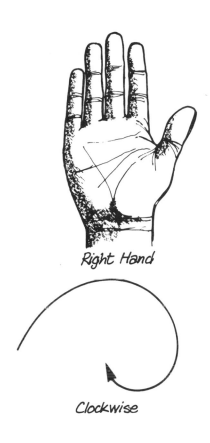

Right Hand

Clockwise

d-Glucose

Twists polarised Light to Left

folk-lore. The very word 'sinister' is derived from the Latin word for 'left' and circling 'widdershins' (anti-clockwise) was an essential part of old magic spells. In the Orient these ideas have an even deeper meaning. The concepts of Yin and Yang—which can be represented by left- and right-handedness—are held to be the complementary opposites which lie at the root of all the natural cycles of change.

This is so close to the viewpoint of modern particle physics that it comes as no surprise to learn that it was two Chinese physicists, Yang and Lee, who first examined the mirror symmetry of inter-actions between the new particles. It seemed impossible to conceive of an event which could only take place right-handedly in space and the Law of Conservation of Parity affirmed that the left-handed, mirror image, of this event should be equally probable. In Yang and Lee's theory this is no longer held to be true and experiment has confirmed their strange prediction.

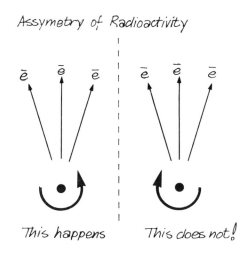

Assymetry of Radioactivity

This happens   This does not!

The old patterns of symmetry are upset when the very lightest particles are involved and Nature itself seems to favour one side of the mirror. Electrons are ejected in only one direction from the gyrating atom and the even lighter, illusive, neutrino seems to be able to remove left-handedness from a particle event and escape at the speed of light, spinning anti-clockwise as it goes!

The Incredible Neutrino

Do neutrinos really exist or are they just a face-saving device invented by physicists to protect their long-established laws of conservation? According to theory neutrinos should be minute even by atomic standards, electrically neutral, have zero rest mass, and travel through either space or matter with the speed of light! The chances of detecting in a lifetime even one of the billions that pass through our bodies each minute are remote indeed. It was calculated that an average neutrino might penetrate *ten thousand million million miles* of solid substance without contact with a single nucleus. Since such an unimaginable thickness of matter is to be found nowhere in the Universe it is little wonder that the neutrino, and its shadow the anti-neutrino, were considered no more than 'ghost particles' for many years. Yet, in 1956, astonishing as it may seem, clear experimental evidence of their existence was obtained. The brilliance of such an achievement, in the face of the overwhelming odds against it, is quite breathtaking!

A powerful nuclear reactor was used as a source of these particles since the neutrons it produced were expected to decay into protons, electrons, and anti-neutrinos within a few minutes of their formation by fission.

Of the estimated million-million-million anti-neutrinos generated per second by the reactor, a large tank of water might be expected on average to capture one every twenty minutes. The lucky hydrogen nucleus would then turn back into a neutron and emit one positron. Such rare events were actually detected even when the water was heavily shielded against all other radiation. As a final check the nuclear reactor was then switched off and 70 less events were recorded per day than with it working. Such delicate and sparse

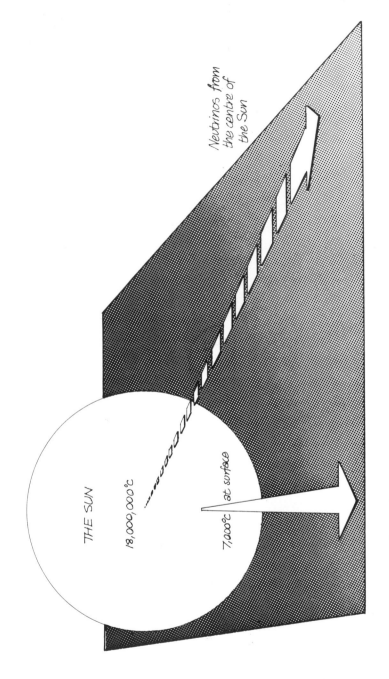

results are taken as proof positive of the existence of this spinning wisp of a particle.

Billions of neutrinos should also be produced in the very centre of the sun where the essential, heat-generating, fusion reaction takes place at temperatures of more than ten million degrees centigrade. Since visible light comes to us only from the cooler exterior of the sun a new, probing science of 'neutrino astronomy' has been hopefully begun in the last few years. Deep underground in disused mines scientists have set up their unwieldy 'neutrino traps' where other radiation cannot penetrate. Here, paradoxically, a mile below the surface of the earth, they hope to probe the secrets of the central core of the sun itself!

Two families of particles

With the experimental proof of the neutrino's existence the family of the very lightest particles became recognisable and, perhaps, complete. Indeed no new members have been added to it for the last thirty years and they stand separate from the new heavier particles with their own pattern and rules. They are called the *leptons* and comprise eight members:

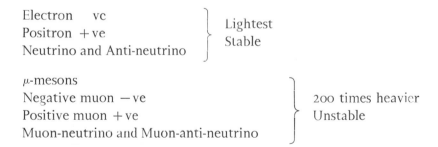

Electron   ve
Positron +ve   Lightest
Neutrino and Anti-neutrino   Stable

μ-mesons
Negative muon −ve   200 times heavier
Positive muon +ve   Unstable
Muon-neutrino and Muon-anti-neutrino

In addition to their vanishingly small mass they have in common three distinct properties:

1 Their number remains unchanged in every reaction. For each additional lepton created an anti-lepton is always formed with which

it could be annihilated; as though the Universe can only tolerate an exact grand total of their members for all time.

2 They react with a very weak force. Transformations in which they occur are correspondingly slow by atomic standards. The quantum of the Weak Force is unknown and its nature is still a complete mystery.

3 Parity is never conserved in their reactions. These happenings have a basic asymmetry being either right-handed or left-handed but never both. The reason for this inherent lopsidedness of action is also quite unknown.

◎ ◎ ◎ ◎

The family of the heaviest particles is much larger, newer, and seems more complex. Their number stands at 64 as far as I know, but may well be greater still, and all except the familiar proton are unstable—many lasting less than a millionth of a millionth of a second after their formation. They are known collectively as the *baryons* and again they exhibit three general characteristics:

1 Their number, like that of the leptons, remains unchanged in the Universe. This grand cosmological number is probably the same as that for the leptons—inconceivably vast, but finite.

2 They 'feel' each other by the Strong Force; this is the reason for their extreme instability—strong forces beget sudden beginnings and abrupt ends!

3 Some members of this group, the heavier ones, can carry a new sort of charge. It is not electric and is still referred to as 'strangeness' or 'hypercharge'.

◎ ◎ ◎ ◎

The alarming proliferation of these strange baryons in the late 1950s led to new methods of classifying them into smaller groups with related symmetries of their new properties. Space itself was

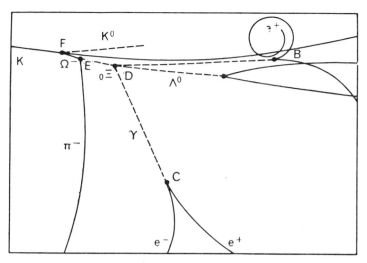

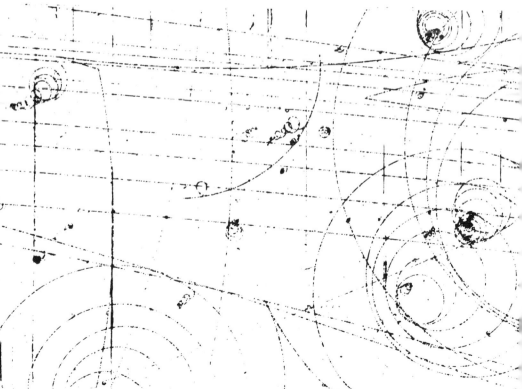

Six 'events' in a Bubble Chamber and the discovery of [Ω⁻] the predicted particle.

treated like a resonance chamber which could react to certain combinations of the Conservation Laws for possible particles, like a hollow jar can resound to certain permitted frequencies. Attempts are being made to sort out the particles into families of eight or ten members linked together like the harmonics of a musical note and in 1964 one of these theories was tested in the time-honoured way.

A prediction could be made that yet another 'strange' particle could exist transitorily, perhaps for a bare ten-trillionth of a second and it was named $\Omega-$ (omega minus). Later the same year this particle was detected with all the properties which had been forecast for it. This mathematical system is called SU_3 or, more poetically, 'The Eightfold Way'. The number eight refers to the possible combinations of conservation symmetries but, like so much in recent physics, an admiration for Eastern philosophy inspired the imagery. It refers to the teaching of the Buddha:

> Now this, O monks, is noble truth that leads to the cessation of pain: this is the noble Eightfold Way: namely, right views, right intention, right speech, right action, right loving, right effort, right mindfulness, right concentration.

More complex systems of symmetry have since been evolved which promise further successes in codifying the baryons.

In between the two classes of heavy and light particles lie the mesons, instruments and carriers of the force between the baryons. Since the family characteristic which distinguishes the particles is no longer mass, spin, charge, or energy but only the nature of the force by which they are created and destroyed, these mesons are now perhaps more fundamental than any of the more 'tangible' particles. Even the one stable and familiar baryon, the proton (hydrogen nucleus), has been found to have a fine structure of a core surrounded by a pulsating cloud of mesons. Perhaps the same is true of all the baryons. We do know that mesons can exist within protons and neutrons in various excited forms of the Strong Force ready to leap into any adjacent baryon attracting, transmuting, and destroying as they move!

The label 'elementary particle' has now gone out of favour. None of the host of baryons can really qualify for that title but, at the same time, the mesons have acquired a new significance. Just as the channels of political communication in the twentieth century can create crises as well as pass on information so, in the study of new particles, the physicists have learnt to see the exchange of mesons in a similar role. In the search for the ultimate units of substantial matter science is faced with a changing panorama of creation and annihilation in which only the interconnecting forces remain recognisable. Whether the next real advance in our understanding will grow from the changing patterns of symmetry which cause these ephemeral particles to materialise so briefly in space, or through a new theory of the 'substance' of the forces of interaction, it is clear only that the subject is still as fascinating as ever and that the long exploration into the basic nature of substance will go on.

Acknowledgements

Acknowledgement is due to the following for kindly allowing their illustrations to be used. The numbers refer to the page on which the illustration appears.

10 British Museum; 25 from *Origins of Alchemy* by J. Lindsay, Muller; 50 Lord Portsmouth; 54, 124 Science Museum, London; 68 Manchester Corporation; 70, 78 Royal Institution, London; 108 from *Science*, Macdonald Illustrated Library, Macdonald; 116 (left) Brian J. Thompson; 116 (right) Radio Corporation of America; 130 Lord Blackett; 152, 154 Washington Atomic Commisison; 163 *Scientific American*, Oct 1963; 171 *Scientific American*, Oct 1964.

Select Reading List

Abell, G. *Exploration of the Universe* (1970)

Asimov, I. *Environments out There* (1968)

B B C Programme Talks. *Einstein, the Man and his Achievement* (1967)

Berlage, H. P. *Origin of the Solar System* (1968)

Bernardini & Fermi. *Galileo and the Scientific Revolution* (1963)

Born, M. *Physics in My Generation* (1970)

Brumbaugh, R. S. *The Philosophers of Greece* (1966)

Crombie, A. C. *Augustine to Galileo, Vol 2* (1969)

Farrington, B. *Science in Antiquity* (1969)

Gamow, G. *Gravity* (in the Science Study Series) (1962)

Gamow, G. *Mr Tompkins* in Paperback (1965)

Harre (Edited by). *Early Seventeenth Century Scientists* (Gilbert, Bacon, Galileo, Kepler, Harvey Van Helmont, Descartes) (1965)

Holmyard, E. J. *Alchemy* (1968)

Hoyle, F. *Galaxies, Nuclei and Quasars* (1966)

Kendall, J. *Michael Faraday* (1955)

Koestler, A. *The Sleepwalkers* (1970)

Koestler, A. *The Watershed* (A biography of Kepler) (1961)

Kopal, Z. *Telescopes in Space* (1968)

Lindsay, J. *The Origins of Alchemy* (1970)

Lochak & Silva, A. *Quanta* (1969)

Lovell, B. *Our Present Knowledge of the Universe* (1967)

Ronchi, V. *The Nature of Light* (1970)

Schatzman, E. L. *The Structure of the Universe* (1968)

Sciama, D. *The Unity of the Universe* (1959)

Index